Networking

Studies in Literature and Science

published in association with the
Society for Literature and Science

Editorial Board

Titles in the series

NETWORKING

Communicating with
Bodies and
Machines in the
Nineteenth Century

Laura Otis

Ann Arbor

THE UNIVERSITY OF MICHIGAN PRESS

2004   2003   2002   2001   4   3   2   1

*A CIP catalog record for this book is available
from the British Library.*

Library of Congress Cataloging-in-Publication Data

Otis, Laura, 1961–
    Networking : communicating with bodies and machines in the
nineteenth century / Laura Otis.
        p.   cm. — (Studies in literature and science)
    Includes bibliographical references and index.
    ISBN 0-472-11213-9 (cloth : alk. paper)
    1. Telecommunication—History—19th century. I. Title. II.
Series.
TK5102.2 .O88   2001
302.2—dc21   2001002844

for Sander

Acknowledgments

Networking is appearing today thanks to the generosity of Hofstra University. Even though I had just taken a one-year leave to complete another book, Hofstra granted me a full-year sabbatical to investigate networks and nerves. This level of support for junior faculty research is almost unheard of, and I deeply appreciate Hofstra's ongoing enthusiasm for my work.

I performed almost all of the research for this book at the University of Chicago libraries, and I am deeply grateful to the people there who helped me locate sources. I am especially indebted to the librarians in Special Collections, who not only brought me good books but encouraged me with their enthusiasm for the project: Jay Satterfield, Krista Ovist, Barbara Gilbert, Jessica Westphal, and Debra Levine. I also did some valuable work at the Philadelphia College of Physicians, where Charles Greifenstein, Director of Historical Collections, continues to be a great help.

I would also like to thank all the members of the University of Chicago's British Romantic and Victorian workshop for their input on the project. I am especially grateful to Sam Baker, Saree Makdisi, Larry Rothfield, and William Weaver for their thoughts on nature vs. culture and to Laura Demanski for her valuable ideas on Henry James. I am particularly grateful, too, to Françoise Meltzer, who offered me a home in the University of Chicago Comparative Literature Program while I was researching and writing this book.

I would like to thank my colleagues this year at the *Max Planck Institut für Wissenschaftsgeschichte* in Berlin for their thoughts on the project: Cornelius Borck, Sven Dierig, Jean Paul Gaudillière, Hennig Schmidgen, Friedrich Steinle, and Ohad Parnes. I am especially grateful to Michael Hagner and Hans-Jörg Rheinberger for making a place for me at the Max Planck Institut and introducing me, as a fellow ex-scientist, to the realm of *Wissenschaftsgeschichte*. I appreciate the help I have received from the librarians at the Max Planck Institut, particularly Ellen Garske and Ulrike Burgdorf who located books and scanned the illustrations.

I am also indebted to the many scholars who have offered feedback on this project at conferences, particularly Katherine Hayles, whose work has always been an inspiration; Jill Galvan; Timothy Lenoir; Richard Menke; Sid Perkowitz; and David Porush. Thanks to Susan Squier, Carol Colatrella, Hugh Crawford, and many others who have kept the Society for Literature and Science running, I have had wonderful opportunities to exchange ideas with other interdisciplinary scholars.

I owe my deepest gratitude for *Networking*, though, to Sander Gilman, my former advisor and now most supportive and inspiring friend. Thank you, Sander, for your endless help to a neurobiologist who wanted to study literature.

Quotations from Mark Twain's stories "Mental Telegraphy" and "From the London Times of 1904" are taken from *The Science Fiction of Mark Twain,* edited by David Ketterer. © 1984 David Ketterer (Hamden, Conn.: Archon Books/The Shoe String Press, Inc.) and are being reprinted by permission.

Contents

Introduction

> But our Grub Street of today is quite a different place: it is
> supplied with telegraphic communication, it knows what
> literary fare is in demand in every part of the world.
> —George Gissing, 1891

What Is Communication?

To communicate is to share, to impart, to touch. If the right sort of connection exists, one can communicate a thought, a disease, or even the Spirit of God. Until the mid-nineteenth century, "communication" referred equally to the movements of information and of physical goods (Carey 16). For this reason, it could mean physical touching or even sexual intercourse. Over the years, our changing ways of communicating have shaped the way we see the world. Not just our messages but our *ways of transmitting them* have suggested what knowledge must be. James W. Carey defines communication as "a symbolic process whereby reality is produced, maintained, repaired, and transformed" (23). As Carey implies, communication is inseparable from culture. Rather than offering ways to discuss a preexisting reality, it constructs the reality that we know.

In a conversation with engineer Werner von Siemens, explorer Alexander von Humboldt called telegraphy *Gedankendrahtung* (wiring thoughts). Having begun his career studying animal electricity, Humboldt saw the cables of the 1840s as nerves transmitting the impulses of a society. Over one hundred and fifty years later, we might still ask what happens to thoughts when they are "wired." Are "wired thoughts" different from those spoken or shouted? Does the medium change the message, and can a "message" be distinguished, abstracted from the medium that transmits it?

If the message and its medium are inseparable, as Humboldt's word suggests, then communications systems alter the way we think, especially the way we think about ourselves. To describe the way information technologies can shape our self-concept, Friedrich Kittler has adopted a term coined by Daniel Paul Schreber, a psychotic who believed that "divine nerve rays" placed him in communication with God. According to Kittler, changing technologies create *Aufschreibsysteme,* "discourse networks," or more literally, systems for writing things down. Supplementing or replacing the body's communications system, these mechanisms suggest how our own bodies work. Communications technologies "structure seeing," Allucquère Rosanne Stone has proposed, because "they act on the systems—social, cultural, neurological—by which we make meaning" (167–68).

Since the late 1840s, electronic communications networks have changed the way we see our bodies, our neighbors, and the world. For a century and a half, these networks have suggested webs, leading their users to think as though they were part of a net. Between 1845 and 1895, the development of the telegraph transformed people's understanding of communication and, with it, their notion of their relation to others. As the telegraph affected language, Carey argues, it "changed the forms of social relations mediated by language" (210). The telegraph became "a thing to think with," shaping the thoughts that it wired (Carey 204).

In the twenty-first century, we still imagine knowledge in terms of our systems for acquiring and transmitting it, and we view the network as the quintessential communications system. We speak of networks of computers, of professional contacts, even of nerves within our bodies. Networks of highways facilitate our movement, and telecommunications nets inform us about the world when we cannot explore it for ourselves. The image of the worldwide web, however, did not begin with the computer. Emerging from studies of nervous and electromagnetic transmissions, the web has been upheld for two centuries as nature's own apparatus for transmitting information. Images of bodily communications nets have inspired us to build technological ones, and images of technological ones have inspired us to see them in the body.

In his Foreword to Kittler's *Discourse Networks,* David E. Wellbery declares that "in its nervous system, the body itself is a medial apparatus" (xiv). He means, of course, that the nervous system is *like* an electronic medium—or does he? If they perform the same functions, are nerves *like* cables or are they identical, members of the same functional category? Metaphors elide likeness, masking a key epistemological link. But what is the epistemological value of metaphor? What does one gain by saying that one thing is like another?

Physicist and physiologist Hermann von Helmholtz, who measured the velocity of the nerve impulse, recognized the importance of this question for scientists. In the opening of his theoretical physics textbook, he challenges readers by asking, "What is likeness?" When we compare two phenomena, he maintains, it is essential to specify the kind of likeness we are observing, watching for tendencies that differ while noting those that coincide. In comparisons of organic and technological communications systems, it could be objected that nerves are alive and thus inherently different from any sort of technological apparatus. Since the early nineteenth century, though, drawing a distinction between organic and technological systems has grown increasingly problematic. No mind can be abstracted from the information system that feeds it, and rather than passively receiving data, the mind controls the circuits that monitor its own environment. If a "medium" is a substance that allows communication, then nerves are undeniably a "medial apparatus." They perform the same functions as technological communications devices, and they can be studied with the same scientific methods.

From the time that investigators first began studying the nervous system, they have described it in terms of contemporary technologies. In the seventeenth and eighteenth centuries, it was a hydraulic system; in the nineteenth, a telegraph net; and in the twentieth, a cybernetic web. What has been the effect of these comparisons? How have these metaphors shaped the way that we think? Have scientists studying nerves simply used these comparisons to convey their findings to laypeople, or have the technological metaphors affected their decisions about which phenomena to study and what experiments to perform?

George Lakoff and Mark Johnson have proposed that metaphors are much more than rhetorical devices for conveying complex ideas. As we form the associations they invite us to make, we do not just learn how to speak and write—we learn how to think. Scientists use metaphors not just to communicate their ideas to the public, as a kind of "dumbing down," but to express their ideas to each other. I agree strongly with historian of science Timothy Lenoir that neuroscientists' technological metaphors reveal the way they understand the nervous system. Nineteenth-century physiologists' instruments inspired their models of nerve-impulse transmission, and from the beginning, the equipment they used to study the living system shaped their understanding of the way nerves work (Lenoir, "Models and Instruments" 1994). The telegraph, for instance, suggests uninterrupted lines of transmission and may have led a generation of anatomists to see a physically continuous nerve net (Dierig 54).

While Lenoir and others have demonstrated how technology has affected neuroscience, I offer new evidence that physiological studies have inspired communications engineers as well. Like technological networks, bodies transmit information. Using electrical signals, they inform us of events in the world and allow us to survive by responding to environmental changes. In 1851, German physiologist Emil DuBois-Reymond called the nervous system a *Vorbild* (model or ideal) for the telegraph. Like technological webs, he saw, organic structures suggested ways to transmit information and organize space.

Norbert Wiener, the founder of cybernetics, based the new science of communication and control upon analogy. As he saw it, "animals, humans, and machines can be treated as equivalent cybernetic systems" (Hayles "How We Became Posthuman" 93). The language of cybernetics must always be analogical, he maintained, since we lack direct access to people's perceptions and can interpret their descriptions only by drawing on our own, similar experiences. By focusing on the likeness of living and technological information systems, Wiener hoped to build knowledge that reflected the structure of thought itself (Hayles "How We Became Posthuman" 92–98).

"When Galvani was making dead frogs dance on the table," wrote historian John Francis in 1851, "he was preparing for one of the most important inventions of recent times" (1: 277). Offering this remark in a popular history of British railways, Francis suggested that the telegraph grew out of studies of organic communications systems. Samuel Morse compared his telegraph lines to nerves, and Alessandro Volta, whose battery powered Morse's communications device, modeled his Voltaic pile on the electric organ of a fish. Since the eighteenth century, scientists designing communications systems have been inspired by the structures of living bodies.

Metaphors comparing living and technological networks, however, have never been restricted to scientists. Just as scientists' metaphors have shaped their own understanding of the nervous system, they have affected that of the public at large. Metaphors are "catchy," among the most infectious associations language offers. The fact that John Francis, a popular historian with no background in physiology, compared telegraph lines to nerves suggests how widespread such comparisons were. A cultural channel for transmitting metaphors, moreover, always involves movement in both directions. In the nineteenth century, descriptions of nerves as wires passed not just from scientists to the public but from the public to scientists.

N. Katherine Hayles has proposed that a "feedback loop" connects science with literature and culture (*Chaos Bound* xiv). In fascinating studies, she and other scholars have illustrated how some of the principles of twentieth-century physics emerged simultaneously in contemporary fiction. Cultural changes challenge writers in all fields with the same quandaries of subjectivity, space, and time. While investigators follow the guidelines of their respective disciplines, they express themselves in a common language, with all of the metaphorical associations that it offers them. Despite the widening gap that technology is creating between science and other fields, writers communicate their infectious images just as they always have.

But how do these transmissions occur? How does information move along the feedback loop between science and culture? What drives the circulation, and what are its limits? Why do certain metaphors propagate themselves endlessly, and why do others die out? To analyze the exchanges between literature and science, I have made a case study of communication itself. In recent years, many scholars have shown how scientists and creative writers use common strategies when they compare computers to brains and brains to computers. With the rapid development of artificial intelligence studies, the idea of the body as a machine has become much more than a metaphor as traditionally defined. While for most people living brains are still distinct from technological information systems, the revelations of neuroscientists and engineers are turning likeness into identity.

Organic systems have been compared to technological ones since ancient times, and the notion of a technological nerve network is much older than the twentieth century. Because communications systems and their representations have developed so rapidly since Galvani's day, I have chosen a particular network to study: that of the telegraph. The telephone became available in 1877 and deserves attention in any study of nineteenth-century communications networks, but I saw that I could only follow the complex exchanges among neuroscientists, engineers, and creative writers if I confined my study to a single telecommunications system during a given period. Beginning with Galvani's twitching frog legs, I have explored writers' images of nerves and telegraph lines from the late eighteenth until the late nineteenth centuries, from the days when scientists first proposed using electricity to "transmit intelligence" to the days when they began transmitting thoughts without wires. It is my hope that this work will supplement that of scholars studying twentieth-century comparisons of computers to brains. The ability to see the central nervous system as a computer grew out of nineteenth-century physiologists' visions of nerves as telegraphs, and understanding the earlier scientists' motives should show us the implications of our own analogies.

In exploring the ways that culture has shaped science, I have granted a special role to literature. To study nonscientists' understandings of communication, I have examined three canonical works (*Middlemarch, Dracula,* and "In the Cage") and a variety of telegraph literature. From a disciplinary perspective, literature might be seen as being as far away from science as one can get. Surely if fiction writers were depicting nets and comparing nerves to telegraphs, such images must have been very widespread. I have selected creative writers, however, not just for the ways that they differ from scientists but for characteristics that they share with them. Like scientists, fiction writers—both "high" and popular—struggle to communicate images only they can see, trying to make these visions as real for their readers as for their creators. To make their perceptions convincing to others, both scientists and novelists rely upon metaphor. Since they must relate to a variety of readers, they seize upon any ready analogies culture has to offer. At the same time, they forge their own metaphors, which then enter the cultural store. To understand how nineteenth-century people thought about communication, it is essential to read the works of scientists and novelists in parallel. Although they lived and worked quite differently, they faced the same challenge to communicate and answered it with cultural knowledge and creativity.

On a cultural level, what makes people think in terms of nets? Why did representations of organic and technological communications networks emerge when they did, and what assumptions about space and subjectivity might have fostered their development? Comparing Charles Darwin's metaphors to those of nineteenth-century fiction writers, Gillian Beer observes that "web imagery is to be found everywhere in Victorian writing. It is as common among scientists and philosophers as it is among poets and novelists" (*Darwin's Plots* 156). A web, of course, is not a net or a network, although in many European languages, their meanings overlap. Two French words, "*toile*" and "*réseau*," and two German words, "Gewebe" and "Netz," correspond approximately to the English words "web" and "network," respectively. As in English, "*toile*" and "Gewebe" denote a structure of woven threads, especially a spider's web. "*Réseau*" and "Netz" suggest an artificial structure. It is tempting to draw a distinction between webs as organic and nets as artificial structures, except that for millennia, people have been weaving both. Beer believes that for most Victorians, the word "web" suggested woven cloth as well as a spider's web of concentric circles. Certainly the rapid growth of the textile industry in the early nineteenth century would have encouraged writers to express relationships in terms of "webs." In the same way, the development of the telegraph and the railways promoted images of networks.

In 1911, parapsychologist William Barrett wrote that the invention of the wireless telegraph vindicated the idea of telepathy. According to Barrett,

> hostility to a new idea arises largely from its being unrelated to existing knowledge. As soon as we see, or think we see, some relation or resemblance to what we already know, hostility of mind changes to hospitality, and we have no further doubt of the truth of the new idea. It is not so much *evidence*

that convinces men of something entirely foreign to their habit of thought, as the discovery of a *link* between the new and the old. (*Psychical Research* 108, original emphasis)

While Barrett's hypothesis of thought transference has not been substantiated, he makes an essential point about the ways ideas are transmitted. Fifty years before Thomas Kuhn, he argues that for scientists as well as laypeople, analogy is often more persuasive than puzzling new facts. Certainly technological discoveries have led scientists to see how nerves are linked and to build more efficient communications systems. Equally important, however, has been the ability to see the connection—to see the relevance of the technological network to the organic communications system.

In *Membranes: Metaphors of Invasion in Literature, Science, and Politics* (1999), I studied how changing understandings of personal and national identity encouraged people of the 1830s to see living things as associations of independent units. In the 1880s, this concept of boundedness helped people to associate diseases with invasive microbes violating individual and national borders. During these same years, however, many neuroanatomists saw the nervous system as a continuous net. In 1887, when some neuroscientists proposed that the body's communications system consisted of independent cells, bitter debates arose between "neuronists" and "reticularists." Clearly, the notions of boundedness and continuity coexisted in time, leading to conflicting visions of the body. In nineteenth-century Western culture, a discourse celebrating individuality collided with an ideology of connectedness, and the interference pattern they created can be seen in scientists' and novelists' representations of communications systems.

Beginning with the problem of individual perception, I open this exploration of metaphoric exchange by studying some prominent physiologists' understandings of nerve impulses: first, as electrical signals and second, as representational signs. Hermann von Helmholtz and Emil DuBois-Reymond, the nineteenth-century scientists who revealed the most about the ways nerves worked, drew upon electrical, magnetic, and telegraphic discoveries to build their instruments and to construct their models of nerves (Lenoir, "Models and Instruments" 1994). Helmholtz, the first scientist to measure the velocity of nerve impulses, compared the body's signals to words in a language. Both, he believed, were representations offering relative but never absolute knowledge of the world. Charles Babbage, who designed a calculating engine to relieve weary brains, shared Helmholtz's interest in sign systems and developed a new mechanical notation to represent the movements of the machine's parts. Both the physiologist studying nerves and the mathematician creating a mechanical brain were aware of the metaphorical nature of language and its unstable connection to physical reality. Like philosopher Friedrich Nietzsche, who called nerve impulses "metaphors" of external events, Babbage and Helmholtz suggested that we could know the world only through metaphorical links.

In the second chapter, I study the life cycle of a particularly hardy metaphor. Here I move from physiology to anatomy, exploring the rise and fall of the nerve-net paradigm and the arguments of its advocates and opponents. Both the neuronists (who believed that neurons were independent cells) and the reticularists (who believed that neurons merged into a net) were excellent scientists and used similar staining techniques. Ironically, those who saw individual cells and those who saw a network accused each other of the same scientific sins: of allowing preconceived ideas about function to distort their ideas about structure and of simply not observing carefully enough. Italian anatomist Camillo Golgi, one of the strongest defenders of the neural network, believed that a communications system could not function without physical continuity. However, Spanish neurobiologist Santiago Ramón y Cajal, whose work overthrew the reticular paradigm, believed that a structural model based on physical merging ignored the nervous system's capacity to grow and change. Philosopher and physiologist George Henry Lewes upheld the neural net for moral reasons, proposing that like a body, a society functioned because all of its elements were in communication.

In the third chapter, I continue studying how network metaphors promoted social and moral philosophies. Here, though, I explore a creative writer's pattern of images, examining the web's incarnations in George Eliot's *Middlemarch*. As knowledgeable as her partner, Lewes, about physics and physiology, Eliot shared his view that organic connections offered the best moral model for social bonds. Because Eliot upheld the railway as a communications system fostering the development of sympathetic ties, I examine the growth of this modern circulatory network in the 1830s and 1840s. In *The Railway Journey*, Wolfgang Schivelbusch describes a nineteenth-century ideology of circulation that sees the transportation demanded by industrial capitalism as healthy and beneficent (195). While Eliot's novel does not celebrate capitalism, it does represent all forms of communication in a positive light and warn about the results of isolation. For Eliot, who depicts a small town's confused response to the railways, awareness of one's position in a larger social organism allows one to act as a more responsible individual.

If the railways offered nineteenth-century societies a circulatory system, the telegraph offered them nerves. As James W. Carey has observed, the relationship between the telegraph and the railways became "an entrance gate for the organic metaphors that dominated nineteenth-century thought" (215). Until the late 1840s, the railways were a communications network, but once the telegraph became operational, the movement of information could be distinguished from the movement of a society's lifeblood (16). In the fourth chapter, I study the development of telegraphy from the mid-eighteenth through the mid-nineteenth centuries, exploring its designers' reliance on living communications systems. Many of the scientists who developed the telegraph got their start studying organic "receivers." Charles Wheatstone and Wilhelm Weber were acoustics experts, Karl August Steinheil was an optician, and William Cooke spent years learning how to make wax anatomical models for medical students. When telegraphers described their work, they depicted their bodies as continuous with their machines. Their stories about telegraphy show that, like the scientists, they understood their affinity with their keys and wires.

In many respects stories by and about telegraph operators challenge widely held notions about the telegraph. As Helmholtz describes, technological communications networks shared the shortcomings of organic systems, offering a mere representation of the world. Like nerves, telegraph wires delivered information that could never be confirmed, and like the language of the nerves, that of the telegraph could lie. Examining two fictional representations of telegraphy, Ella Cheever Thayer's *Wired Love* and Henry James's "In the Cage," I compare the young female protagonists' dissatisfactions with their communications webs. While both operators enjoy the power and freedom that telegraphy brings, they eventually recognize how little knowledge it actually provides. Epistemologically considered, the telegraph was a flawed communications device, offering only signs of a distant, unknown reality.

The development of wireless telegraphy did not make the telegraph a more reliable "nerve net," nor did it slow comparisons between organic and technological communication. In a final chapter, I explore arguments for the direct transmission of thoughts—some of them offered two decades before Marconi developed the "wireless." If both nervous systems and technological networks conveyed impulses as energy fluctuations, reasoned scientists of the late 1870s, then why should not minds someday transfer thoughts to one another directly? Investigators for the British Society for Psychical Research, founded in 1882, insisted that if currents could be induced from a distance, then the same should be true for thoughts. Mark Twain parodied these arguments in his ironic tales "Mental Telegraphy" (1891) and "From the *London Times* of 1904," and Bram Stoker incorporated them into his novel *Dracula* (1897). Pitting cutting-edge communications technology against an organic communications system, Stoker juxtaposes Dracula's telepathy with his hunters' telegraphy. Just as the vampire hunters communicate through electronic impulses transmitted by wires, Dracula issues orders directly to his servants' unconscious minds. At the same time, Stoker challenges people's confidence in their technological nerve nets. For much of the novel, Dracula eludes his modern hunters, and his telepathic web works better than their telegraphic one. While Stoker relies on contemporary comparisons of organic and technological thought transmission, his novel raises serious doubts about whether artificial information systems will ever surpass living ones.

Like our own comparisons of brains to computers, these early alignments of bodies and technologies altered people's sense of identity. The tendency to see a communications device as a continuation of one's own nervous system developed in the nineteenth century, not in the twentieth. As early as the 1870s, self-conscious telegraphers felt themselves merging with their networks, describing the transmission of signals from their brains, through their fingers, onto their keys, and then on down the line. Then, as now, sending electronic messages challenged the traditional notion of a bounded, delimited individual. As part of a network, one is defined through one's connections to others, with an identity that is better represented by a vector diagram than by a bounded cell. In nineteenth-century accounts of the interfaces between living and technological information systems, one can hear our own anxiety about where "we" end and our networks begin.

Chapter 1
The Language of the Nerves

In a popular lecture in 1851, German physiologist Emil DuBois-Reymond (1818–96) proposed that

> the wonder of our time, electrical telegraphy, was long ago modeled in the animal machine. But the similarity between the two apparatus, the nervous system and the electric telegraph, has a much deeper foundation. It is more than similarity; it is a kinship between the two, an agreement not merely of the effects, but also perhaps of the causes.[1]

In 1851, the telegraph and the nervous system appeared to be doing the same things, and for the same reasons. Their common purpose was the transmission of information, and they both conveyed information as alterations in electrical signals. By calling the nervous system a model for the telegraph, DuBois-Reymond suggested that organic communications systems offered solutions to problems encountered by brand-new technological ones.

In the same decade, however, French physiologist Claude Bernard (1813–78) remained skeptical about the epistemological value of metaphor. From the classical notion of nervous fluid to the more recent ones of animal spirits and animal electricity, he pointed out, people's theories about the way the nervous system worked had consisted largely of a series of analogies, "the expression of a way of seeing meant to explain the facts" [*destinée à expliquer les faits*] (1: 3). Priding himself on his empiricism, Bernard mistrusted analogy as a means of constructing knowledge.[2] Does one know more or less about something if one asserts that it is like something else? What exactly is the relationship between metaphor and knowledge?

In his textbook on theoretical physics, German physicist and physiologist Hermann von Helmholtz (1821–94) pondered "the concept of likeness" (*Gleichheit*), considering the question as essential for science as it was for philosophy. One could claim that something was "like" something else, he concluded, only in a restricted and relative sense: only by comparing both objects to a third one and only by specifying the parameter by which the two were compared (*Vorlesungen über theoretische Physik* 26–28). In his own scientific writing, Helmholtz struggled to stand by this finding, always following a claim for likeness with more specific statements about the ways in which two concepts were alike. By constructing metaphors, one was not building objective knowledge, but one was creating productive thought.

Traditionally, many investigators have shared Bernard's concern that claims of likeness will change their perceptions of physiological systems. For the same reason, however, some scholars of language are much more optimistic about the epistemological possibilities of metaphor. George Lakoff and Mark Johnson assert that the role of metaphor in scientific thinking shows how metaphors reflect and influence what we see and do. "Formal scientific theories are attempts to consistently extend a set of ontological and structural metaphors," they argue (220). This "extension," though, is just what worried Bernard. Metaphors provoke and give birth to new images; by establishing and reinforcing connections, they encourage us to see in new ways. While Bernard is correct that assertions of likeness alter the way we see, Lakoff and Johnson are equally correct that "much of cultural change arises from the introduction of new metaphorical concepts and the loss of old ones" (145). Alterations in the way we see can be extremely productive.

One of the most intriguing cases through which to study the epistemological role of metaphor is that of comparisons between organic and technological communications systems. Throughout the nineteenth century, scientists' electrophysiological understanding of the nervous system closely paralleled technological knowledge that allowed for the construction of telegraph networks. When one reads the proposals of nineteenth-century engineers, one begins to suspect that communication in the body and in society can only be understood in terms of each other. In her history of mesmerism, Alison Winter shows the power of a "common vocabulary": nineteenth-century people thought of physical forces, mental states, and social changes in the same ways because they used the same words to describe them (*Mesmerized* 18, 276). At the same time, of course, they applied terms like "electric" to all three concepts because they thought about them in similar ways. The vocabulary of electronic communications shaped thought, which in turn shaped the language of communications.

Norbert Wiener has claimed that "it is certainly true that the social system is an organization like the individual, that it is bound together by a system of communication" (24). Can one say, though, that comparisons of communication in bodies and machines actually contributed to our knowledge of the nervous system and early telecommunications networks? What experiments—illustrative of Lakoff's and Johnson's idea—did such comparisons inspire, and what thoughts—illustrative of Bernard's concern—did they prevent?

The development of transportation systems, communications systems, and neurophysiology in Europe and the United States coincide in time to a remarkable degree. All three, of course, are closely related to the explosive growth of mechanized industry, the first two for obvious reasons, the third for less obvious ones. The invention of the spinning jenny in 1770 and the rapid growth of the weaving industry—particularly in England—fed the societies in which early neuroscientists lived and thus could never have been far from their minds. The image of the intricate web in which each point was interconnected was as relevant to economics as it was to biology and physics.

By the mid-nineteenth century, the ubiquitous steam engine and the growing understanding of thermodynamics suggested that bodies really *were* like machines, so that comparisons between the two were not just aesthetically provocative but scientifically accurate. Many of the same scientists whose experiments shaped our current understanding of thermodynamic laws, physicists like Hermann von Helmholtz and Julius Robert Mayer (1814–78), were also physiologists interested in the body's energy relations. Explaining "The Interaction of Natural Forces" to the public in 1854, Helmholtz wrote that "the animal body . . . does not differ from the steam-engine as regards the manner in which it obtains heat and force, but . . . in the manner in which the force gained is to be made use of" (*Science and Culture* 37). By specifying differences as well as similarities, Helmholtz encouraged his audience to think further about the two concepts he was comparing.

From the late eighteenth century onward, scientists studying organic and technological communications systems continually inspired one another. As Friedrich Kittler argues in *Discourse Networks*, nineteenth-century media for writing and communication—the telegraph, the typewriter, and later the phonograph and telephone—affected not just the way people wrote and communicated but the way they perceived their own minds and bodies. At the same time, people's understanding of organic structures shaped the communications devices they built. As can be seen in DuBois-Reymond's identification of the nervous system as a "model" for the telegraph, physiologists and physicists drew upon one another's representations of communications networks. Even if their images were not adopted by other scientists directly, they were often seized upon by popular culture, so that the individuals studying telegraphs and nerves stimulated one another in a complex "feedback loop."[3] To follow the exchange of images among nineteenth-century scientists describing communication is to enter a complex circuit of thought—a system of coils, cross-links, and loopings in which a fluctuation at any point instantly becomes a property of the entire system.

The Metaphoric Circuit

Since ancient times, Claude Bernard observed, people have attempted to understand how the nervous system works by comparing it to something else (1: 3–5). Observing the brain's ventricles filled with cerebrospinal fluid, the earliest anatomists envisioned the nerves as a kind of circulatory system, drawing inferences about their structure and function by comparing them to a system whose structure and function were more obvious. Both systems seemed to involve branching channels that grew progressively finer as they approached the body's surface, and the hypothesis developed that nerves were hollow tubes for conducting "nervous fluid" from the brain to the muscles and extremities. This hydraulic model persisted from classical times until the early nineteenth century, and even in 1833, anatomist Christian Ehrenberg (1795–1876) wrote that "the cerebrum can obviously be compared to a *capillary vascular system for the nerve fibers*" (Clarke and O'Malley 42, original emphasis).

For seventeenth-century thinkers, nervous fluid and animal spirits were scientific facts and concrete realities. René Descartes (1596–1650) believed that the body consisted of tiny particles in perpetual motion, the most active and rarefied of which reached the brain to produce sensation. These "animal spirits" then flowed through the nerves to produce activity when they reached their destinations in the muscles (Brazier, *Neurophysiology in the Seventeenth and Eighteenth Centuries* 21–22). When the body needed to move, the nervous fluid quite literally "pumped up" the muscles, readying them for action.

Not everyone accepted this hydraulic model. One of the earliest objectors was Antony von Leeuwenhoek (1632–1723), who in 1674 examined a cow's nerves with his microscope and wrote, "I could find no hollowness in them . . . as if they only consisted of the corpuscles of the Brain joined together" (Clarke and O'Malley 31). In the mid-eighteenth century, Albrecht von Haller (1708–77) changed the way scientists viewed the relationship between nerves and muscles by positing the fiber as the basic unit of both systems. Because he could see no more reasonable hypothesis for nervous conduction, however, he retained the notion of nervous fluid. Despite his vision of nerves as fibrous, he rejected the only competing hypothesis, that of the nerves as cords for transmitting vibrations, because the nerves seemed soft and watery and lacked the necessary tautness. "The nervous fibers cannot possibly tremulate in an elastic manner," he wrote, "neither at their origin, nor where they are tense," because they were "firmly tied to the solid parts by means of the cellular fabric" (220).

Based on his careful observations, Haller proposed that all the body's fine structures comprised a continuous "net-like cellular substance" (12). "None of the cellular fabric," he asserted, "is excepted from this communication" (18). Studying the way nerves separated into fine threads to descend into the muscles, Haller inferred that in such an arrangement, the body and its controlling mind must be intimately related. After more than a decade of experiments begun in the 1730s, Haller distinguished irritability, a general property of living tissue but especially of muscle, from sensibility, a property unique to the nerves. While nerves differed from other tissues in their ability to transmit impressions, they worked so closely with the rest of the body that one could only view them as part of a continuous whole. "The fibers of the brain are continuous with those of the nerves, so as to form one extended and open continuation," wrote Haller (219). Such observations led him to see the body and mind as interdependent. Rather than believing in a "soul" or "spirit" distinct from organic matter, he insisted that "we must reckon the mind to be changed, when any change happens to the body" (214). This perspective would make possible a whole host of studies about the body's relationship with electricity.

In 1747, taking Haller's theories as a point of departure, Julien Offray de la Mettrie (1709–51) proposed that people functioned like machines. By "machine," the French doctor meant a device that regulated itself and could respond to situations automatically. While he occasionally compared the human body to a clock, he had no visions of internal cogs or gears, nor did he argue that the body ran on electricity. Instead, he built his argument slowly and carefully on Haller's evidence for the interdependence of mind and body. Descartes had already suggested that animals were machines, of sorts, and la Mettrie asserted that the difference between people and animals had been greatly exaggerated. People's vaunted ability to reason had not made them any less ferocious to one another (la Mettrie 172). If there was any difference between people and animals, it lay in people's ability to manipulate systems of signs—those of language and of mathematics. It simply made no sense to set up an opposition between matter and spirit when the "human machine" so clearly showed what matter could do on its own.

As the second pillar of his argument, la Mettrie incorporated Haller's idea of the fiber as the body's elementary unit. Because all fibers were irritable, it was reasonable to see each one as alive, inherently capable of motion. Taken as a whole, the body could be viewed as an "assembly of springs [ressorts]" (la Mettrie 186). It required no consciousness separate from its fibers to set it in motion, for its control mechanism was already an integral part of the "machine" (la Mettrie 182). What most clearly identified the body as a machine was its organization, each part deriving its function and identity not from its own distinct nature but from its position in a system of interconnected parts. To demonstrate the body's ability to function independent of any conscious control, la Mettrie listed a great variety of automatic responses. Physiological arousal in response to thoughts of a beautiful woman occurred through "exchanges [*commerce*] and a kind of sympathy of the muscles with the imagination" (la Mettrie 183). The body, "a living image of perpetual motion," required no soul to regulate or initiate its movements (la Mettrie 154). Thought, when properly understood, was a fundamental quality of organized matter, just like impenetrability or movement—or electricity (la Mettrie 192).

In 1747, la Mettrie's identification of electricity as a fundamental property of organized matter lacked experimental support, but it was not as unusual a claim as one might expect. The electric fish, which stuns its prey with a potent shock, had been known since classical times. When scientists and the public began to experiment with electricity in the mid-eighteenth century, people were quick to link the shocks transmitted in parlor games to the fast-moving "fluid" in the nerves. As soon as electricity became generally known, it was seized upon both as a potential "nervous agent" and as a means of communication.

Luigi Galvani (1737–98) produced the first widely publicized evidence that the nerves and muscles of animals used their own intrinsic electricity. An obstetrician, Galvani had particular reason to be interested in muscle contractions (Pera 64). In the early 1780s, he began experiments to test a hypothesis that had long appealed to him: the idea that animals could move because of the electricity they contained. Galvani stimulated dissected frogs' legs with electricity, trying to specify the conditions under which he could produce muscle contraction. His goal was to determine the relation of this "animal electricity" to artificially generated electricity that behaved according to known rules. If it could be stored in Leyden jars, blocked by insulators, and had positive and negative components, then it must bear a close affinity to the "common" electricity studied in laboratories (Pera xxiii–xxiv). Particularly exciting to Galvani and the European public was the result that when his preparations were strung along an insulated wire during a thunderstorm, "as often as the lightning broke out . . . all the muscles fell into violent and multiple contractions" (Galvani 36). The image of Galvani's frog legs twitching merrily when strung along an iron railing was widely disseminated in popular culture.[4]

Based on his results, Galvani proposed in *De Viribus Electricitatis* (1791) that all animals contained electricity in most parts of their bodies, especially

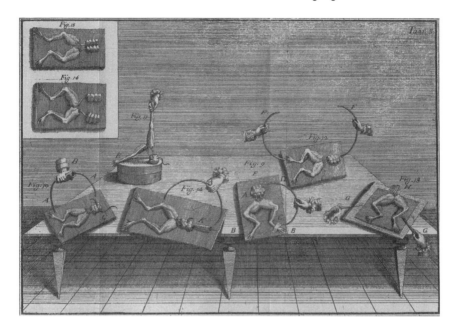

Luigi Galvani's organo-technical circuits for studying animal
electricity in the 1780s. The use of disembodied hands to
represent experimenters is an iconographic convention of
eighteenth-century electronics. (From Luigi Galvani, *Commentary on the Effects of Electricity on Muscular Motion*,
trans. Margaret Glover Foley, Norwalk, CT: Burndy Library,
1954, plate 3).

the muscles and nerves. While he criticized the traditional animal spirits
model of nervous action, he conceived of electricity as a fluid, as did most
other scientists of his day. Just as he took pains to ensure that electrical fluid
did not "find a way for itself through the cracks" of his Leyden jars, he wondered how "the electric fluid should be contained within [nerves] and not be
permitted to escape and diffuse to neighboring parts" (33, 64). Organic and
technological systems, Galvani saw, needed to solve the same problems, and
he pictured the nerves' insulation in terms of the materials available in his
laboratory.

Envisioning the nervous system as a living version of his eighteenth-century electrical apparatus, Galvani proposed that

> it would perhaps be a not inept hypothesis . . . which should compare a muscle fiber to a small Leyden jar, or other similar electric body, charged with two
> opposite kinds of electricity; but should liken the nerve to the conductor, and
> therefore compare the whole muscle with an assemblage of Leyden jars. (61)

Generally, he referred to his experimental set-up as a "circuit" even when the conducting pathway included nerves, muscles, or the hands of experimenters. A cautious scientist, Galvani was acutely aware of his use of analogy. While he performed experiment after experiment to prove that animal electricity resembled the better-known common electricity, he hesitated to identify the two, writing that his hypothesis rested both on experimental results and on "factors of analogy and reason" (52).[5] An empiricist, Galvani mistrusted metaphor as a means of furthering knowledge. He relied upon it, however, to shore up his arguments when he realized that he could never produce irrefutable experimental evidence for animal electricity with the time and technology available to him.[6] Ultimately, his metaphors and diagrams communicated his hypothesis so successfully that the concept of animal electricity became known not just to scientists but to the public at large.

Not all scientists accepted Galvani's interpretations of his results. Alessandro Volta (1745–1827), who initially believed in animal electricity, became Galvani's strongest opponent in debates during the 1790s. According to Volta, the electricity Galvani had observed was real enough, but it had not resulted from any intrinsic properties of animal tissue. Any two dissimilar metals connected by a moist conducting substance would allow the same current to flow, and Galvani's animal electricity was actually an artifact produced by the two metal electrodes used to detect it. "Most of the galvanic phenomena," Volta declared, "had nothing whatever to do with animal electricity" (DuBois-Reymond, *On Animal Electricity* 10). Volta's investigation of the current that flowed between different pairs of metals eventually led to the discovery of the battery (then known as the "Voltaic pile"). The tremendous technological value of this discovery persuaded most scientists to accept his interpretation over Galvani's.[7]

But Volta, who denied the existence of animal electricity, had modeled his battery on an organic structure, the electric organ of the torpedo. Known since antiquity, this electric fish constituted one of the strongest pieces of evidence for the animal electricity hypothesis and had been intensely studied since the 1770s (Pera 60–63). In his letter of 20 March, 1800, describing his Voltaic pile to the Royal Society, Volta called it an "*artificial electric organ, . . . which being at the bottom the same as the natural organ of the torpedo, resembles it also in its form*" (Pera 158, original emphasis). In constructing the battery, Volta went out of his way to give it an organic look, and he suggested to the Royal Society how future versions might be modified to look even more like the natural organ: "These cylinders would have a pretty good resemblance to the electric eel; and, to have a better resemblance to it even externally, they might be joined together by pliable metallic wires or screw-springs, and then covered with a skin terminated by a head and tail properly formed" (Mertens 307). As Joost Mertens points out, Volta always emphasized the "physiological effects" of metallic electricity when he described his inventions because he knew that such references would interest a broad audience (311). These organic images provided Volta with much more than an opportunity for popularization; they inspired his technological innovations. Just as the nervous system would suggest how to build a telegraph, the electric fish suggested how to build a battery.

Following Volta's first challenges, a scramble ensued to substantiate Galvani's idea, for the enormous possibilities of electricity made the concept of animal electricity highly appealing. To prove that electricity was inherent in muscle tissue, one would need to produce contractions either with two identical metal electrodes or in the absence of any metal. Galvani himself struggled to do so but died in 1798 before he could present skeptics with conclusive evidence. His nephew Giovanni Aldini (1762–1834) did not help matters by traveling through Europe and offering animal electricity demonstrations. Aldini, who had a shady reputation and occasionally took credit for other scientists' work, became famous for electrifying the heads of recently decapitated criminals, producing grimaces and hideous contortions (Brazier, *Neurophysiology in the Nineteenth Century* 2; Clarke and O'Malley 184; Pera xii–xiii).

The greatest challenge to Volta and the best defense of Galvani—apart from Galvani's own ongoing experiments—came from the German scientist Alexander von Humboldt (1769–1859). By bending and excising portions of the nerves, he produced muscle contractions in the absence of stimulating electrodes. Humboldt "established circuits" in his dissected preparations by exposing different regions of nerves and muscles to points with which they were not normally in contact (DuBois-Reymond, *On Animal Electricity* 16–17; Brazier, *Neurophysiology in the Nineteenth Century* 12). His experiments strongly suggested that animal movements involved forces independent of those generated by dissimilar metals.

Like Humboldt, Carlo Matteuci (1811–65) observed in the 1830s that when an amputated frog's leg was placed in contact with a leg undergoing contractions, it would contract as well. As Galvani had noted, the frog preparation itself proved to be "the most delicate electrometer yet discovered" (Pupilli xi). Using this organic "device," Matteuci discovered an ongoing current (*la corrente propria*) in frog muscle, which he could detect with particular clarity in cases of injury. In an 1838 study of the electric fish, he called the currents he detected in its nerves "the greatest analogy that we have between the unknown force in nerves and that of electricity" (Brazier,

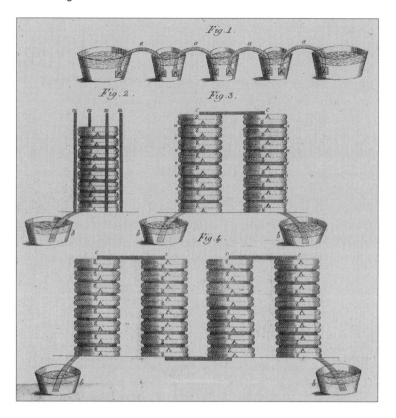

Alessandro Volta's battery. (From Alessandro Volta,
Collezione dell' opere del cavaliere Conte Alessandro Volta
[Firenze: G. Piatti, 1816], vol. 3, plate 2).

Neurophysiology in the Nineteenth Century 31). Despite the success of
Volta's battery, debates revived about the relationship between electrical
forces studied in laboratories and those observed in living bodies. Some
readers took Matteuci's results to mean that electricity underlay all human
interactions (Winter, *Mesmerized* 277). Was animal electricity *like* ordinary
electricity, its name and existence based on a metaphor, or was it identical
with the electricity studied by physicists? Did it advance or obstruct knowledge of the nervous system to compare the nervous impulse to an electrical
signal?

In 1841, the highly respected physiologist Johannes Müller (1801–58)
presented Emil DuBois-Reymond with Matteuci's results and asked that he
establish, once and for all, whether the nervous principle was electrical in

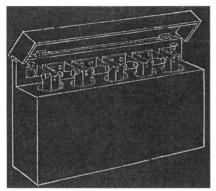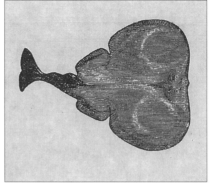

A pocket battery and a *Raia torpedo*. Almost eighty years after the experiments of Galvani and Volta, Prescott juxtaposed the battery with the organic structure that inspired it. (From George B. Prescott, *History, Theory, and Practice of the Electric Telegraph* [Boston: Ticknor and Fields, 1860], 35.)

nature. Müller himself had his doubts. "When the actions of common electricity . . . became more fully known," he wrote, "physiologists imagined that the action of the nerves was rendered more intelligible by comparing them with electric apparatus" (1: 682–83). He believed, however, that "the galvanic fluid . . . is electric in its nature, and altogether different from the nervous principle, acting merely as a stimulus of the nervous force. . . . it is now certain that the phenomena produced in animals by galvanism are not due to an animal electricity" (1: 683). Several inescapable facts illustrated a fundamental difference between neural and electrical signals: (1) a ligated (tied or crushed) nerve could conduct electricity but could not transmit the nervous principle; (2) many other types of stimuli besides electricity could excite nerves, giving rise to the nervous principle; and (3) other moist animal tissues could conduct electricity as well as nervous tissue, if not better. To call the nervous principle electricity, Müller warned, was to call likeness identity.

Writing in the late 1830s, as Samuel Morse (1791–1872), William Cooke (1806–79), and others perfected their electrical telegraphs, Müller insisted that "to speak . . . of an electric current in the nerves, is to use quite as symbolical an expression as if we compared the action of the nervous principle with light or magnetism. Of the nature of the nervous principle we are as ignorant as of the nature of light and electricity" (1: 689). For Müller, who trained a generation of rigorous empiricists, an electrical signal was merely an analogy for the nervous principle, a symbol of the impulse used to communicate in bodies. Experiment, not metaphor, would illustrate the way that organic communications systems worked.

In the 1840s, physiologists found it difficult to specify what the material basis of the nervous principle might be. Isaac Newton (1642–1727) had proposed in his *Opticks* (1704) that thoughts and sensations were conveyed as vibrations of the ether in the nerves, and for almost 140 years, scientists more or less adhered to this model (Brazier, *Neurophysiology in the Seventeenth and Eighteenth Centuries* 91–92).[8] What one *could* know about the nervous agent was its properties, just as one could know the properties of electricity without comprehending its actual nature. Emil DuBois-Reymond shared Müller's belief in experimentation, and he devoted his life to developing techniques for studying electrical activity in nerves and muscles.

Both early neurophysiologists and the developers of the telegraph relied on the same discoveries about electricity and magnetism to design their apparatus, particularly the Leyden jar (1746) and later the Voltaic pile (1800). Even more significant were Hans Christian Oersted's (1777–1851) finding in 1820 that the flow of electrical current through a coil of wire could deflect a magnetic needle and Michael Faraday's (1791–1867) discovery in 1831 that a magnetic fluctuation occurred whenever an electrical circuit was established or broken. These studies of electromagnetic induction allowed scientists studying both organic and artificially constructed circuits to detect the presence of electricity (Gillispie, *Edge of Objectivity* 441–42; Standage 23–24). Scientists' frequent comparisons of nervous elements to Leyden jars, batteries, conductors, and galvanometers suggest that their familiarity with electrical circuits affected not just the way they performed their experiments but the way they conceived of the nervous system itself (Lenoir, "Models and Instruments").

Michael Faraday's theories of electromagnetic induction proved crucial for both telecommunications and electrophysiology. A pioneer in the investigation of electricity and magnetism, Faraday studied the ways that magnets and electrical currents interacted. Self-taught and possessing almost no knowledge of mathematics, he worked out the principles of induction by visualizing the ways in which electrical and magnetic forces influenced one another. While he drew on the work of Oersted and André Marie Ampère (1775–1836), Faraday is largely responsible for nineteenth-century scientists' understanding of electrical and magnetic forces (Gillispie, *Edge of Objectivity* 435–58).

In his *Experimental Researches on Electricity* (1831–55), Faraday defined induction as "the power which electrical currents may possess of inducing any particular state upon matter in their immediate neighborhood" (4). One such state was the "electrotonic" one, a condition that "relate[d] to the particles, and not to the mass, of the wire or substance under induction" (Faraday 20). This condition of the molecules was not itself a current, explained Faraday, but it could be considered the equivalent of a current because currents were measurably increased or decreased when a conductor's electrotonic state was changing.

Like Galvani and DuBois-Reymond, Faraday regarded animal electricity as a legitimate electrical force, claiming that "electricity, whatever may be its source, is identical in its nature" (102). Throughout his career, Faraday maintained a vivid interest in the electrical and "magnetic" phenomena in living bodies. In 1838 he observed mesmeric experiments at the University College Hospital in London, and in 1853 he designed a device to show that table-turning was caused by witnesses' unconscious muscular movements (Winter, *Mesmerized* 52, 266). Faraday collaborated with Charles Wheatstone (1802–75), a developer of the electrical telegraph, to see whether the electricity in the *gymnotus* was identical to that of nonliving sources (Winter, *Mesmerized* 39, 358). While studying the electric fish, he argued that he was not violating the "principle of life," only exploring the forces it produced (Winter, *Mesmerized* 298). Faraday believed in the equivalence not only of electrical forces but of magnetic, gravitational, and chemical ones as well. His vivid descriptions of electrical and magnetic fields convinced many readers of these equivalencies before they could be demonstrated mathematically (Gillispie, *Edge of Objectivity* 441).

In 1848, Emil DuBois-Reymond dedicated his textbook *On Animal Electricity* to Faraday and based his model for electrical excitation in nerves and muscles on Faraday's descriptions of induction in electric circuits.[9] Drawing a "direct analogy" between organic and electrical systems, DuBois-Reymond even borrowed Faraday's term in order to describe an altered, excited state of the nerve: the "electrotonic state" (Lenoir, "Models and Instruments" 12).[10]

Using increasingly sophisticated stimulating electrodes and galvanometers, DuBois-Reymond observed distinct changes in the electrical activity of nerves and muscles. Under normal conditions, the longitudinal surface (the outside) of a muscle or nerve was positive relative to the transverse (inner) surface. During stimulation and contraction, however, this imbalance was either diminished or reversed. In his highly influential textbook, *On Animal Electricity,* he asserted that "muscles and nerves . . . are endowed during life with an electromotive power . . . [that] acts according to a definite law"

(210). Like Galvani, he sought to demonstrate the affinity of this animal electricity to the electricity that physicists studied by proving that the two followed the same rules. DuBois-Reymond believed that his experimental results of the 1840s "restor[ed] to life . . . [a] hundred-year-old dream . . . the identity of the nerve substance with electricity" (Clarke and Jacyna 183).

Inspired by Faraday, DuBois-Reymond proposed the first experimentally supported mechanism for nerve-impulse conduction. The neuroplasm, he believed, consisted of a series of "electro-motive molecules," elementary particles with one positive and one negative side. These particles could change their orientation according to their electrochemical environment (DuBois-Reymond, "On the Time Required" 132). When a stimulus excited the nerve, it was not creating or destroying any electromotive elements; it was polarizing the particles, inducing a temporary electrical change by flipping their orientation. In his eyes, the fundamental particles of organic tissue were simply following the same physical and chemical laws as those of the metals undergoing changes in an electrical circuit.

Like Galvani, DuBois-Reymond described the body's electrical activity in terms of the equipment available in his laboratory. Using his instruments as metaphorical vehicles, he relied upon metaphor to envision what these devices could not yet show him: how nerves and muscles communicated impulses at the molecular level. By comparing the electrical activity in muscles, nerves, batteries, and coils of wire, he was better able to understand the "network of interconnections" that his data offered (Lenoir, "Models and Instruments" 15).

So obvious were the similarities between transmitting signals over nerves and transmitting them over wires that DuBois-Reymond used the metaphor throughout his career to illustrate key aspects of the body's communications network. He opened a lecture of 1868 by declaring that:

> just as little as telegraph-wires, do the nerves betray by any external symptom that any or what news is speeding along them; and, like those wires, in order to be fit for service, they must be entire. But, unlike those wires, they do not, once cut, recover their conducting power when their ends are caused to meet again. ("On the Time Required" 97)

In both the organic and technological systems, a signal "meant" nothing until interpreted by the appropriate receiving organ.

The inability of nerves to resume activity when spliced, however, indicated a key difference between organic and technological communications systems. These systems were analogous, but not identical. Later in the same lecture, DuBois-Reymond stated that the nervous principle "must necessarily be something material, [but] not . . . the electricity such as it moves along a telegraph-wire" (99). He continued to refer to the telegraph as the most useful analog for the nervous system, however, because it so clearly indicated differences as well as similarities.

Real Time

While nervous impulses resembled telegraphic signals, a number of factors still distinguished the two. To understand the relationship between animal electricity and the electricity transmitted along wires, physiologists realized that they would have to focus on these differences. One appeared to be the velocities at which the respective signals moved, although physiologists lacked any definite information about the rate of nerve-impulse transmission. Johannes Müller had written that "we shall probably never attain the power of measuring the velocity of nervous action; for we have not the opportunity of comparing its propagation through immense space, as we have in the case of light" (1: 729). In 1850, however, Helmholtz was able to measure it accurately by adapting the techniques of ballistics and telegraphy for use in the physiological laboratory (Dubois-Reymond, "On the Time Required" 102–3; Lenoir, "Helmholtz" 185–88).

No one illustrates the kinship of nineteenth-century physics and physiology better than Helmholtz, who made significant contributions to both fields. Like DuBois-Reymond, Helmholtz did his best to explain human sensory systems in terms of physical laws, overcoming experimental or theoretical obstacles through his ability to reason from analogy. By his own account, he particularly excelled at applying the techniques of one field to the problems of another (*Science and Culture* 387). Some of the instruments Helmholtz designed for scanning the eye, for instance, were based on devices for astronomical observation. As a physicist, Helmholtz advanced people's understanding of thermodynamics, proposing the Law of the Conservation of Energy in 1847. He is better known, however, for his achievements in physiology and medicine: the measurement of the velocity of the nerve impulse (1850); the invention of the opthalmoscope for exploring the retinal surface (1850); and his still-accepted theory of color vision. An outstanding writer and speaker as well as experimentalist and theoretician, Helmholtz liked to explain scientific discoveries to the public. In his widely read popular essays on energy, optics, and acoustics, he followed the same strategy he did in the laboratory, configuring the problems of perception in terms of physics and those of physics in terms of the body.

Between 1845 and 1855, Helmholtz worked actively with members of the Berlin Physical Society, scientists from diverse fields who shared his interests in imaging, graphic displays, and the measurement of very small time intervals (Lenoir, "Helmholtz" 188). Among these was Werner von Siemens (1816–92), who starting in 1849 designed Prussia's telegraph network. Trained as a ballistics expert in the Prussian army, Siemens had developed a new technique in 1845 for measuring the velocity of mortar shells (Lenoir, "Helmholtz" 187). When Helmholtz studied Siemens's instrumental solutions to the problems of ballistics and telegraphy, he began to see the nervous system and sensory organs as a "media apparatus" (Lenoir, "Helmholtz" 185).

Since 1843, Helmholtz had been exploring the relations among chemical, mechanical, and thermal changes in organic systems. A close friend of DuBois-Reymond, he believed that the electrical changes his colleague was observing were related to these transformations as well. Helmholtz's famous essay "On the Conservation of Force [*Kraft*]" (1847), which stresses the interchangeability of natural forces, emerged from his studies of muscle action and animal heat (Olesko and Holmes 66–67). In 1849, hoping to study the time course of muscle contraction in greater detail, Helmholtz adapted French physicist Claude Pouillet's (1790–1868) method of using a galvanometer to measure minute time intervals (Olesko and Holmes 84–85).

To study muscle contractions, Helmholtz designed an electrical circuit that included the sciatic nerve and gastrocnemius muscle of a frog. At the instant the muscle contracted, it lifted a weight, breaking the circuit. In ballistics, an analogous setup had enabled experimenters to calculate high velocities by measuring the brief interval during which current had flowed (DuBois-Reymond, "On the Time Required" 102). In the organic system, Helmholtz discovered—quite by accident—that when he stimulated the nerve instead of stimulating the muscle directly, the time between the stimulus and the contraction varied significantly according to where he stimulated along the nerve. Intrigued, Helmholtz halted his studies of muscle action, realizing that he had found a way to measure how rapidly a nerve could transmit signals (Olesko and Holmes 87).

In a series of trials, Helmholtz stimulated the nerve and then measured the length of time required to produce a contraction in the muscle. Gradually, he moved the stimulating electrode further and further away from the muscle and studied the way in which this time interval varied with distance. Once one subtracted a certain minimal latency, which occurred even when one stimulated directly at the muscle, it became apparent that the impulse

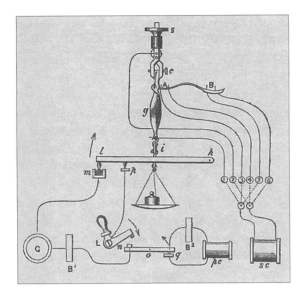

Hermann von Helmholtz's organo-technical circuit for determining the velocity of nerve impulses along a frog's sciatic nerve (*AB*), with some minor modifications by Emil DuBois-Reymond. Upon receiving a stimulus from an induction coil (*sc*), the muscle (*g*) contracts, lifting a weight (*i*) and breaking the circuit at point *n*. The velocity is measured by stimulating the nerve at different points between *A* and *B* and comparing the lengths of time between the delivery of the stimulus and the interruption of current flow. (From Emil DuBois-Reymond, "On the Time Required for the Transmission of Volition and Sensation through the Nerves," in *Croonian Lectures on Matter and Force*, ed. Henry Bence Jones [London: Churchill, 1868] 104.)

was traveling along the sciatic nerve at a rate of 26.4 meters per second. In 1834, Charles Wheatstone, inventor of the British needle telegraph, had calculated the velocity of electricity to be over 250,000 miles per second. German physicist Rudolph Kohlrausch, who later repeated Helmholtz's experiment in sensory nerves as "an exercise in understanding notions of error and precision," measured the velocity of electrical signals along a wire and got a figure of 3.1×10^8 meters per second (Olesko and Holmes 106; Gillispie, *Edge of Objectivity* 472–73). Clearly, the living and the telegraphic communications systems were not functioning in the same way.

When Helmholtz and DuBois-Reymond wrote up Helmholtz's results, they presented the delay between stimulus and response in his frog "circuit" as his most significant finding.[11] It was this interval of a thousandth of a second that had led him to perform the experiments, and it was this interval that he had been struggling to measure with accuracy (Olesko and Holmes 88). "I have found," reads Helmholtz's opening sentence, "that a measurable interval of time elapses during which the impulse . . . is transmitted to the sciatic nerve's entry point in the gastrocnemius muscle."[12] The corresponding report to the French Academy of Sciences, translated by DuBois-Reymond and presented by Alexander von Humboldt, called this interval "un espace de temps," reflecting both the experiment's and the human mind's tendency to express space and time in terms of each other ("Note sur la vitesse de propagation" 204). In the detailed article that followed his initial report, Helmholtz specified that muscle contraction involved three stages: a latency period between the time the stimulus arrives and contraction begins, the contraction itself, and a return to the resting state. In the translation, DuBois-Reymond called this latency period "le temps perdu," a characteristic unique to organic communications systems (Helmholtz, "Note sur la vitesse de propagation" 263). When discussing the implications of Helmholtz's achievement in 1868, DuBois-Reymond introduced a term that we, in the cybernetic age, believe ourselves to have invented: "real time" ("On the Time Required" 99).[13] Because of the way the nervous system worked, there would always be a significant "space of time" between the events that stimulated neural signals and living creatures' perceptions of these events.

Intrigued by Helmholtz's finding that the nerve impulse traveled with a finite velocity, a number of scientists attempted to measure the time required for more complex operations in the brain. Dutch physiologist John Jacob de Jaager gave subjects mild electric shocks and compared the times required for response when they knew or did not know where the shock was going to occur. Adolphe Hirsch, a Swiss astronomer, asked subjects to respond to different syllables and colors and found, as de Jaager had, that they took .15 to .25 seconds longer when they did not know in advance what the sound or color would be. "It . . . appears," DuBois-Reymond concluded, that 'quick as thought' is, after all, not so very quick" ("On the Time Required" 126–29, original emphasis).

While "le temps perdu" seemed to separate living beings from the physical world around them, Helmholtz felt that it permitted a consistent relative, if not absolute knowledge. As long as "le temps perdu" remained constant and as long as one focused on the times of perceived events *relative to each other,* one could use the language of the nerves to construct a reasonably faithful representation of the world outside. One had only to look at the telegraph to see the plausibility of this model. In the telegraph, as in the nervous system, what produced meaning was not the signals themselves but the receiving apparatus. "In the net-work of telegraphs," Helmholtz wrote, "we find everywhere the same copper or iron wires carrying the same kind of movement, a stream of electricity, but producing the most different results in the various stations according to the auxiliary apparatus with which they are connected" (*Science and Culture* 150). For Helmholtz, the principles of telegraphy revealed the way the body processed information: in both systems, indistinguishable impulses created by very different causes became meaningful only when they were received and interpreted.

The Mechanics of the Mind

Two decades before Helmholtz and DuBois-Reymond looked to technology to understand how the body transmitted information, British mathematician Charles Babbage (1791–1871) designed a machine that worked like a brain and could perform functions once associated exclusively with the human mind. Throughout his long, turbulent career, Babbage struggled to make physical and mental labor more efficient. In his critical studies of manufacturing, he approached bodies and machines in the same way, studying patterns of movement and seeking the simplest arrangements of parts that could produce a desired motion. Deeply interested in communications, cryptography, language, and thought, Babbage sought simpler, more appropriate patterns of signs to describe mental and mechanical operations. While he never believed the mind could be replaced by a machine, Babbage thought that some of its activities were purely mechanical and could be better performed by steam. Only when he began building his calculating engines did he realize the insights they offered into the way that the human nervous system really worked. The better one got to know machines, the better one understood the body and mind. Technology suggested not just what questions to ask about the nervous system but how to perform the experiments and what sort of answers one might expect to find.

The human mind's lamentable capacity for error inspired Babbage to design the Difference Engine in 1820. Going over his human "computers'" incorrect calculations for the Astronomical Society, he remarked to mathematician John Herschel, "I wish to God these calculations had been executed by steam!" (Babbage 2: 15; Buxton 46; Moseley 65). Encouraged by

Herschel, who saw no reason why this could not be done, Babbage began designing a machine that could calculate and print out vast tables of data. By the 1820s, such tables were invaluable not just to astronomers but to the nautical and insurance industries, although unfortunately, as Babbage revealed in one study, "the quantity of errors from carelessness . . . will scarcely be believed" (2: 8). Between 1820 and 1832, as he painstakingly designed his calculating engine, he learned about manufacturing, toolmaking, and mechanical drawing, his original, creative mind allowing him to make lasting contributions to each field.

Babbage's quest for efficiency, simplification, and near-organic unity recurs throughout his explorations of vastly different fields. As early as 1820, he had seen that mental labor, like physical, could be effectively divided. He expressed this perspective best in *The Economy of Machinery and Manufactures* (1832), his most successful work. By the time he published this study, Babbage had examined a great deal of machinery in workshops and factories throughout England and the Continent (Babbage 8: vi; Buxton 65). Manufacturing on a grand scale not only added greatly to human power, he argued, but offered economy of time (Babbage 8: 6). The division of labor was perhaps the most essential principle of manufacturing, and just as dividing physical labor increased efficiency on the factory floor, the division of mental labor would allow managers "to purchase and apply to each process precisely that quantity of skill and knowledge which is required for it." What sense did it make to hire an "accomplished mathematician [to perform] the lowest processes of arithmetic?" (8: 141). By dividing mental labor and assigning the purely "mechanical" tasks to machines, one could eliminate errors, conserve mental energy, and save a great deal of time.

When Babbage describes the manipulation of tools, it becomes clear that he sees the human body as another sensitive instrument the engineer must learn to read. As a boy, Babbage was once so excited by an exhibition of automata that the manufacturer invited him to visit his workshop. Later in life Babbage purchased his favorite, the twelve-inch "silver lady," and displayed it to people visiting his salon (Winter, *Mesmerized* 57). With his collaborator and soul mate, mathematician Ada Lovelace (1815–52), he joked about teaching an automaton to play ticktacktoe (Moseley 35, 158, 200). Like others in his generation, he was impressed from an early age with the degree to which machinery could mimic the body's movements, and these successful imitations of physical activities suggested future successes in reproducing mental ones.

Babbage's view of the body as a sophisticated machine emerges strongly in his *Reflections on the Decline of Science in England* (1830). Here he tries to teach scientists how to observe. It is absolutely essential, he tells his readers, to get to know one's instrument. Describing an experiment in which he and a friend repeatedly tried to stop a watch in the same position, he suggests that the first instruments whose limitations one must understand are one's own hand and the nerves that drive it. "Both the time occupied in causing the extremities of the fingers to obey the volition, as well as the time employed in compressing the flesh before the fingers acted on the stop, appeared to influence the accuracy of our observations," he remarks. "The rapidity of the transmission of the effects of the will depended on the state of fatigue or health of the body" (7: 88). Intrigued by his experiment, Babbage expressed a desire "to compare the rapidity of the transmission of volition in different persons, with the time occupied in obliterating an impression made on one of the senses of the same persons." As an example, he proposed moving a red-hot coal in a circle at different speeds and comparing the velocities at which different observers perceived a continuous orange streak (7: 88). Helmholtz and other physiologists would begin performing similar experiments two decades later.

Like the physiologists who explored the nervous system, Babbage viewed the nerves in terms of mechanics because he spent so much of his life struggling with instrumentation. It is hardly surprising that a mathematician who studied machines like a naturalist (he once described London printing presses as "rich in minutiae") should view the mind and nerves in technological terms (Buxton 65). Still, Babbage never viewed *all* of the mind's activities as mechanical, only some of them. These activities were the ones that his calculating engines were to supersede.

Both Italian mathematician Luigi Federico Menabrea (1809–96), who explained the value of Babbage's Analytical Engine in 1842, and British mathematician Ada Lovelace, who translated and annotated Menabrea's essay, emphasized Babbage's distinction between mechanical operations and higher reason. The mind is working mechanically, Menabrea explained, when it is doing simple arithmetic, when it is "subjected to precise and invariable laws, that are capable of being expressed by means of the operations of matter" (Babbage 3: 93). As a Cambridge mathematician, Babbage knew very well how the drudgery of endless calculations degraded the mind. In his earliest descriptions of the Difference Engine, he wrote of the "intolerable labor and fatiguing monotony" of these "lowest operations of the human intellect" and of the "wearisomeness and disgust, which always attend the monotonous repetition of arithmetical operations" (2: 6, 15). In designing a calculating engine, he sought not a substitute for the human mind but for the error-prone "compositor and the computer" (Babbage 2: 7). The goal was to free the mind from mechanical labor so that it could devote itself to activities no machine could perform.

Perhaps because of the popularity of automata, Babbage and his advocates took great pains to distinguish his engines from the more versatile human brain. When Menabrea described Babbage's Analytical Engine, he reminded readers that "the machine is not a thinking being, but simply an automaton which acts according to the laws imposed upon it" and "executes the conceptions of intelligence" (Babbage 3: 98, 112). By referring to the better-known automata, which could perform patterns of physical motion but could not "decide" which motions to make, Menabrea hoped to convey the engine's ability to perform mental operations without judging which calculations to do. Babbage's appointed biographer, self-educated mathematician Harry Wilmot Buxton, specified that "the machine, although automatic, has no pretension to originate thought; for though it is an automaton, and acts subserviently to any laws which may be impressed upon it, it is nevertheless incapable of discovery, or of dealing with data, where the laws of the problem are not previously known." He reassured readers that "the engine cannot reason, nor does it in any way encroach upon the functions of the human understanding, its office is to execute the conceptions of the mind" (160, 166). Babbage himself saw his engines as pitifully crude in comparison to the living mind. In his *Ninth Bridgewater Treatise* (1837), he admitted that "all such engines must ever be placed at an immeasurable interval below the simplest of nature's works" (9: 5).

The harder Babbage worked on his calculating machines, the deeper was his interest in the human brain and nerves. Freed from boring calculations, the mind could devote itself to worthier pursuits, and none was worthier than "attend[ing] to the progress of its own operations" (Buxton 60). As Babbage struggled to design his Difference Engine, he concentrated as hard as he could on the way his own mind worked, seeking the solutions to technical problems in organic processes. "I was . . . anxious to watch," he wrote, "with all the attention I could command, the progress of the mind, in pursuit of mechanical invention, and to arrest if possible, the perishable traces of its course and to communicate to the world what is of far higher value than the most ingenious machine,—the art by which it was contrived" (Buxton 60). In a moment of insight, Babbage declared in *The Economy of Machinery and Manufactures* that the principles governing the arrangement of factories "are founded on principles of deeper root than may have been supposed, and are capable of being usefully employed in preparing the road to some of the sublimest investigations of the human mind" (8: 135). When one systematically studied ways to "organize" factories, one discovered principles of "deeper root," principles governing the operation of organic structures "designed" to perform analogous tasks. Technology suggested how the mind and body solved problems, and organic systems suggested ways to build better machines.

Even as Babbage and sympathetic mathematicians indicated the key differences between minds and machines, they revealed a great deal about their underlying similarities. Waxing enthusiastic, Buxton forgets himself for a moment in declaring, "the marvelous pulp and fiber of a brain had been substituted by brass and iron, [Babbage] had taught wheel work *to think*, or at least to do the office of thought" (48–49, original emphasis). As Babbage himself stressed, his engines could not think, but their basic organization reflected the nervous system's structure as revealed by anatomists of the 1830s. In attempting to build a machine that could perform complex operations, Babbage always subdivided what might have been an elaborate structure into many simpler ones (Buxton 63; Moseley 78). Because the engines worked according to the "method of differences," there was a "great uniformity" in their parts, and they were organized in terms of repeating units (Buxton 54). One great advantage of this arrangement was that theoretically, one could progressively add more and more units to form ever more powerful compound machines (Buxton 58). When Babbage describes his own efforts to simplify the Difference Engine, he writes as though he were viewing a living animal. As a designer, he seeks near-organic unity. "Individuals became substitutes for classes," he recalls, "actions intended to operate simultaneously were effected by the same agent, and the connections of the several parts with the moving power becoming more apparent, gave to the design a unity in which it had appeared deficient" (Buxton 64). Even though the engines performed only a small subset of familiar mental activities, they looked and worked a lot like the human brain.

When one considers the structure of the Difference and Analytical Engines, their resemblance to living "computers" becomes even more apparent. Babbage knew little or nothing about anatomy or physiology, and before the articulation of cell theory in the late 1830s, the organization of the nervous system was poorly understood.[14] By his own admission, however, Babbage thought intensely about how the brain must work, and the organic quality of his designs owes itself to these thought experiments. Babbage's machines were not simple calculators. In their design, they reflected the basic principles of analysis. They relied on the ideas that complex calculations could be reduced to the four basic operations of arithmetic and that

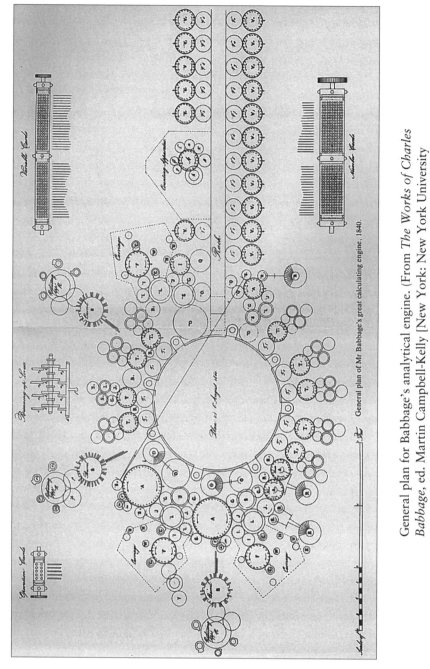

General plan for Babbage's analytical engine. (From *The Works of Charles Babbage*, ed. Martin Campbell-Kelly [New York: New York University Press, 1989], vol. 3.)

complex functions could be represented using the coefficients of terms in a series (Babbage 3: 112).

While the Difference Engine, which generated tables of data through successive additions and subtractions, was "merely the expression of one particular theorem of analysis," the Analytical Engine, which Babbage designed in the 1830s, exceeded it in power and utility because of its simpler, more general design (Buxton 156; Moseley 137–38). Ada Lovelace called it "the material expression of any indefinite function of any degree of generality" (Babbage 3: 115). Like the Difference Engine, which Babbage designed in the early 1820s, the Analytical Engine consisted of columns of toothed discs, the "variable columns." While the earlier engine had had only seven columns, the Analytical Engine would have upwards of two hundred. Each disc on a column was numbered, from zero through nine, and could rotate independently, the first column representing the ones, the second the tens, the third the hundreds, and so forth for as many powers of ten as one desired (Babbage 3: 99–100, 128–29). These variable columns constituted a sort of short-term memory for original figures and intermediate values until all calculations were complete. Menabrea called them a "store" and Lovelace a "storehouse" for numbers (Babbage 3: 105, 128). The actual operations were performed by the "mill," which could be placed in a "state" of addition, subtraction, multiplication, or division. As Lovelace indicated, the most ingenious aspect of Babbage's design—one that eluded many mathematicians—was the independence maintained between the variables and the operations to be performed (Babbage 3: 116–17). It was from this independence that the machine drew its wonderful versatility, for it could work not just with any mathematical function but with an object as specific as "the fundamental relations of fixed sounds in the science of harmony" (Babbage 3: 118).

To avoid errors in entering and recording data, Babbage tried to make his calculating engines as independent of their human operators as possible (Buxton 53). Inspired by Joseph Marie Jacquard's (1752–1834) automatic loom, he used punch cards rather than human hands to give the Analytical Engine its directions. In the French loom, each group of threads that were to work together to create a design was connected to a common lever. These levers were then grouped so that when a card was passed over them at constant speed, each one was depressed except when a hole directly over its position allowed it to be raised. In the loom, this ingenious arrangement allowed complex patterns to be woven at a greater speed and greatly reduced cost. Some patterns required as many as twenty thousand cards (Babbage 3: 101–3; Buxton 163–65; Moseley 186). In the Analytical

Engine, "variable cards" specified the numbers that were to be acted upon, entering them into the columns, while "operation cards" placed the mill in the appropriate state. According to Buxton, Babbage's use of the cards "enlarged the sphere of [the engine's] capability to an indefinite extent" (163). Babbage had succeeded in designing—though he never succeeded in building—a machine that could analyze complex functions as the human mind did, but with greater accuracy and speed.

It was the abstract, advanced nature of Babbage's thinking, not any chronic "irascibility" on his part, that made it so difficult for the mathematician to convey his ideas to others. Quite possibly, his lifelong interest in communications grew out of his own personal struggles to communicate (Buxton 127; Hyman xi; Moseley 18). Only a few far-seeing mathematicians, such as the Italian Menabrea and Babbage's friend Ada Lovelace, understood the theory behind his calculating machines. Babbage could not have been a poor salesman, however, for he so thoroughly convinced the British government of his Difference Engine's value that it contributed seventy-five hundred pounds toward its design and construction. Curiously, Babbage failed to reach an understanding with the government about much simpler issues, such as whether, in the mid-1830s, he should abandon the Difference Engine and devote himself to the more versatile Analytical Engine (Moseley 100, 139). His work with manufacturing, the post office, the railways, and other scientists and mathematicians showed him how essential efficient communications systems were to industrial societies, and his own frustrating efforts to communicate sparked an ongoing interest in sign systems and the way they worked.

In his *Ninth Bridgewater Treatise,* Babbage offered the printing industry as evidence of people's ability to change the world. "Easy and cheap methods of communicating thought from man to man," he proposed, were what allowed civilization to develop (9: 14). Inventions like the speaking tube contributed to the "economy of time," freeing workers to perform more productive tasks (Babbage 8: 8). From his own experience, Babbage pointed out how essential rapid, open communications were to scientific progress. In *The Decline of Science in England,* he argued that British scientists were falling behind Continental ones because the British isolated themselves within their Royal Society and lacked the Continental scientists' professional network for communicating ideas internationally (7: ix).

Because of these convictions, Babbage devoted considerable time to improving British communications systems, working as a consultant to the post office and railways and writing to the *Times* about the need for a telegraph system. Approaching the postal network as he approached his machines, Babbage took nothing for granted, seeking to simplify and streamline accepted procedures. He suggested a way to transmit letters in small cylinders over short distances and in a systematic study discovered that a uniform charge for postage stamps was more efficient because the cost of transport was insignificant compared to the cost of collection and distribution (Moseley 53–54).

The railways had always intrigued Babbage, and in 1830 he personally attended the opening of the Liverpool-Manchester Railway, one of England's earliest lines. In 1838 Babbage worked as a consultant to the Great Western Railway and designed a "dynamometer car," equipped with a seismic graph system to study the motions of the train. For safety's sake, Babbage recommended, every engine should have a way to monitor its own velocity, and every train should carry a dynamometer to keep track of its motions. While engaged in these applied studies of mechanical motions, he became friends with George Stephenson, England's foremost railway engineer (Moseley 130–31). Babbage enjoyed his work with the railways and was very bitter when petty politics kept him off a committee to investigate railway accidents in 1846 . Eventually, one of his two surviving sons would work as a railway surveyor (Moseley 198–200, 164). Like most designers of the telegraph, Babbage maintained a lifelong interest in communications systems of every sort—both those of society and those of the human body.

Like Babbage's interest in the railways, his interest in telegraphy can be explained partly by his passion for innovative, well-designed machinery. When he visited Italy in 1838 to discuss his Analytical Engine with Continental mathematicians, he fascinated the Italian king with his discussions of electric telegraphy, which was then still largely an idea (Moseley 141–42). Babbage always kept himself apprised of the many ways to transmit information, and during the Crimean War, he wrote to the *Times* that the allies had lost the Battle of Sebastopol because a general had misunderstood a signal. England was ignoring the aids of "a highly advanced state of mechanical science," he admonished, and if its generals did not have an "instant means of communication" it would lose the war. Babbage recommended an occluding telegraph, a system through which signals were transmitted by periodically covering and uncovering a light source. The Russians, he pointed out, were using such a system already (Babbage 5: 85–86). If rapid, accurate communications systems were necessary for science and industry, in war they were a matter of life and death.

Babbage was intrigued by sign systems throughout his life, and like Charles Wheatstone, he had a passion for cipher. As a schoolboy, he excelled at designing and deciphering new systems and could crack a challenger's cipher even if he knew only a few words. For Babbage, working through a secret system of signs let him divert his mind by "overcoming a difficulty for its own sake" (Moseley 40, 129). Deciphering a message, Babbage wrote in 1855, is like picking a lock (another diversion he enjoyed): it can always be done; the question is how much time it will consume and whether it is worth the investment (5: 91). Babbage approached codes as he approached challenging mechanical problems, relying on mathematics to reveal relationships that might not be immediately apparent (Hyman xi). In one systematic study in 1831, he compared the frequencies of occurrence of all twenty-six letters in English, French, German, and Italian and recommended that further studies be made involving two- and three-letter words (Babbage 4: 126–27). Such information would have been vital to telegraphers, for those using Claude Chappe's visual communications system in the 1820s and 1830s were as interested as their successors in maintaining secrecy and sending messages in cipher. Astronomer F. Baily, for whom Babbage once deciphered a letter, observed that "by a laborious and minute examination and comparison of all the parts, [Babbage] at length obtained the key to the alphabet" (Babbage 4: 147).

In language and in sign systems in general, Babbage sought the same virtues as he did in machinery: simplicity, efficiency, and a touch of genius in the design. In an exchange of letters with cipher enthusiast John H. B. Thwaite in 1854, Babbage scornfully dismissed Thwaite's claim that he had discovered a new system based on the "principle of permutation"—a system, Thwaite emphasized repeatedly, that would be ideal for electrical telegraphy. Thwaite's use of successive codes, Babbage remarked, "add[ed] much to the labor but nothing to the security of his cipher" (5: 77). Years of experience had taught him that "it requires a very small exertion of intellect to contrive a very difficult cipher"; the real challenge was to design a cipher that was difficult to crack but easy to read, write, and translate (Babbage 5: 78). With his taste for masterful designs, Babbage refused to take on a new system unless he knew that its creator had deciphered difficult systems himself. In language, as in mechanics, he had no patience for systems in which the designer had not fully thought through the relationships among all of the parts.

So carefully did Babbage consider these relationships that by the time he abandoned the Difference Engine in 1834, the mechanical drawings of its parts occupied ninety boxes. These drawings represented its elements in different positions relative to one another. The problem, of course, was that a drawing could depict each part in only one position at one time, with the result that Babbage had to carry a synthesis of the ninety boxes of drawings in his own head. In a discussion with the Countess of Wilton, Babbage told her that his greatest challenge was keeping in mind "the almost innumerable combinations among all the parts—a number so vast, that no human mind could examine them all" (Moseley 98). He speculated to the Duke of Wellington, a strong supporter of the Difference Engine, that picturing all of its parts in motion must be a great deal like commanding an army, and the general replied that he understood the problem well (Moseley 98). Like controlling soldiers in battle, coordinating the movements of the Difference Engine constituted a challenging problem in dynamics, one that ordinary language was not well equipped to represent. As Babbage struggled to transcribe the engine's motions, he "became aware of the necessity of adopting some mechanical language, by means of symbols, by which his overwrought brain might be relieved" (Buxton 128). The designer's brain was undergoing such stress not because of any inherent inadequacy but because of the inability of written language to record the information he had to "keep in mind."[15]

Babbage solved the problem by inventing a new language—mechanical notation—more appropriate for representing the movements of machines. Worn out with "classifying and grouping the innumerable combinations" of motions, Babbage decided that "the forms of ordinary language were far too diffuse to admit of any expectation of removing the difficulty" (Moseley 98; Babbage 3: 209). Knowing "the vast power which analysis derives from the great condensation of meaning in the language it employs," he decided that "the most favorable path to pursue was to have recourse to the language of signs." For his mechanical notation, Babbage sought a system that would be "simple and expressive, . . . capable of being readily retained in the memory from the proper adaptation of the signs to the circumstances they were intended to represent" (Babbage 3: 210). The designer redesigned language so that the relationship between each signifier and its mechanical signified became more "natural" and "appropriate." Like an ideograph, each signifier avoided representing concepts through sounds and instead simply looked like what it was. In mechanical notation, as described in Babbage's article of 1826, the symbols corresponding to the Difference Engine's parts indicated their velocity, their angular velocity, the direction of their motion, the parts to which they were connected, and the way in which they were connected. By inventing this new system, Babbage hoped he had created "a vast and all-powerful language . . . for the future use of analysis," and he promised that "the signs, if they have been properly chosen, . . . will form as it were a universal language; . . . they will supply the means of writing down at sight even the most complicated machine" (Buxton 163; Babbage 3: 217).

The choice of symbols was all-important, Babbage and his advocates knew, for the symbol that represented a concept affected the way one thought about that concept. Ada Lovelace stressed this issue with particular vehemence when she explained common misunderstandings of the terms "operation" and "variable." What made analysis so difficult for many people to understand, she observed, was that the symbols kept shifting meaning. The symbols of operations were also commonly used to represent their results or just numerical magnitude. "Wherever terms have a shifting meaning," she pointed out, "independent sets of considerations are liable to become complicated together, and reasonings and results are frequently falsified" (Babbage 3: 117–18). Even as he explained that the Analytical Engine could never interpret its own results, Menabrea qualified: "unless indeed this very interpretation be itself susceptible of expression by means of the symbols which the machine employs" (Babbage 3: 112). A significant part of interpretation was translation, and if the machine "spoke" the language into which the results were to be interpreted, then it was interpreting its own results.

Even while they acknowledged that signs affected one's understanding of a concept, Menabrea, Lovelace, and Babbage did not believe that language was identical to thought or that any sign system could represent the internal processes of a mind or machine. The beauty of the Analytical Engine, Lovelace proposed, was that it could "arrange and combine its numerical quantities exactly as if they were *letters* or any other *general* symbols." As it currently functioned, the machine "translat[ed] into *numerical* language general formulae of analysis already known to us. . . . But it would be a mistake to suppose that because its *results* are given in the notation of a more restricted science, its *processes* are also restricted to those of that science" (Babbage 3: 144, original emphasis). In Babbage's engines, the processes performed were independent of the symbols through which their results were eventually represented, and this feature of his machines, perhaps more than any other, reflected his understanding of the human nerves and mind.

Both Babbage and Lovelace were "addicted to parrots." Fascinated by these random generators of language, the mathematicians joked about them in their letters, and Babbage allowed one bird to sit beside him in the evening and tell him when to go to bed (Moseley 203). What automata are to motion, parrots are to language: they reproduce patterns that have been impressed upon them, but they do not design their own patterns or "think" about the best—or worst—time to reproduce the patterns they have "learned." Babbage's keen interest in parrots and automata indicates his common understanding of the way that minds and machines work. For Babbage, thought and language were largely independent, and the symbols emitted by a body or engine had little to do with the processes being carried out.

In an early article describing how he got the idea to perform calculations by steam, Babbage uses a popular eighteenth-century metaphor, writing that "when we have clothed [an idea] with language we appear to have given permanent existence to that which was transient, and we admire what is frequently only a step in the process of generalization as the creation of our own intellect" (2: 16). Applying a word to a concept, he suggests, is a very minor aspect of thought, "dressing up" a notion without making any real contribution to its form. In one of his many speculations about how the mind functioned, Babbage implied that it worked with symbols rather than with ideas directly. These symbols, however, were by no means identical to those used in spoken language. Babbage believed that "where ideas can be accurately expressed by symbols, and their relations to one another, by similar means, the law of operations would seem sufficient to effect the rest. It is pretty certain that in the exercise of our reasoning faculties we deal more largely with the symbols of our ideas and their relations to one another, than with the ideas themselves" (Buxton 154).[16] The mind and nerves had their own system of mechanical notation, and by designing two calculating engines and a new language with which to represent their parts, Babbage was creating a technological version of an organic design. Whether Babbage's work with mechanical systems shaped his view of organic ones or whether his insights about organic systems suggested how to design technological ones will never be fully known. Certainly he knew more about engineering than he did about anatomy or physiology. His understanding of the nerves and their "language," though, is remarkably close to that of Hermann von Helmholtz, who would argue four decades later that nerves transmitted signals having little to do with the events they represented.

The Language of the Nerves

Helmholtz never ceased probing the foundations and limits of human knowledge, and his lifelong interest in perception derived not just from his profound interest in the way the body worked but from his philosophical roots in Kantian epistemology. In the 1830s, his mentor, Johannes Müller, had developed a hypothesis loosely supporting Kant's ideas about how much one could know. Like Kant's philosophy, Müller's hypothesis of specific sense energies professed that the mind relied on fixed structural categories to make sense of the world (Dosch 49; Lenoir, "Eye as Mathemati-

cian" 110–18; Fullinwider 44). Rather than serving as "passive conductors," Müller argued, sensory nerves and organs had a "special sensibility to
certain impressions" (1: 819). For any given sensation, the qualities perceived resulted not from any inherent properties of the external objects
exciting them but from the properties of the sensory apparatus responding
to them. According to Müller, "the sensation of light in the eye . . . is not a
development of the matter of light, but is merely the reaction of the optic
nerve" (1: 673). The same stimulus would be perceived differently depending upon the nerve endings it reached. A ray of sunlight that was warmth to
the skin would be interpreted as light by the eye. Müller concluded that:

> Sensation . . . consists in the communication to the sensorium, not of the
> quality or the state of the external body, but of the condition of the nerves
> themselves, excited by the external cause. We do not feel the knife which
> gives us pain, but the painful state of our nerves produced by it. . . . We com-
> municate . . . with the external world merely by virtue of the states which
> external influences excite in our nerves. (1: 819–20)

For Müller, our sensations and our knowledge reflected not the world but
our limited capacity to perceive it.

Helmholtz agreed with his teacher about the limits of human perception,
arguing that one must never confuse actual phenomena with our experience
of them. "The way in which [colors] appear," he asserted, "depends chiefly
upon the constitution of our nervous system" (*Science and Culture* 167). As
he had shown when assessing the problem of "likeness" in physics,
Helmholtz saw that "a property or quality can never depend upon the
nature of one agent alone, but exists only in relation to, and dependent on,
the nature of some second object, which is acted upon" (*Science and Culture*
168). While Helmholtz for the most part retained Müller's idea of specific
sense energies, he moved further and further away from his teacher's
nativist ideas about perception. By the late 1850s, new discoveries in optics
had convinced Helmholtz that vision, like hearing, was a learned process in
which the brain used the eye as a "measuring device" (Lenoir, "Eye as
Mathematician" 111).[17] Although Helmholtz insisted that all knowledge
was learned and could be traced to sensations, he warned that the "fundamental distinction does not completely depend on the type of external
impression by which the sensation is stimulated; rather, it is determined
completely, solely, and exclusively by the sensory nerve that has been
affected by the impression" (*Science and Culture* 345). In the body, as in the
laboratory, people were limited by the sensitivity of their instruments.

In his autobiographical sketch, Helmholtz described the brain as the experimenter's most valuable tool and asserted that one must understand "the capabilities of our power of thought" exactly as one must understand the capabilities of the telescope or galvanometer one worked with (*Science and Culture* 389). Everywhere in the body, Helmholtz witnessed interactions that suggested well-known mechanized processes, and he pointed out these parallels to his readers to enhance their appreciation of both. "The moving force of the muscle must be at work in [the arm,]" he wrote, " and [the muscles] must obey the nerves, which bring to them orders from the brain. . . . Just so is it with machines" (*Science and Culture* 98). In this comparison he singled out weaving machines, "the work of which rivals that of the spider" (*Science and Culture* 99). Just as the human arm resembled the mechanical weaver, the weaving machine resembled the natural spinner.

Perhaps because of his experience as a laboratory scientist, Helmholtz most frequently compared human organs to manmade instruments when assessing their abilities to detect the phenomena around them. He described the eye as a camera obscura and commented that despite its remarkable powers, one would be very annoyed at an optician who sold one an optical instrument with the same defects. In one of his most intriguing analogies, he compared the inner ear to a piano, writing that in this exquisitely sensitive instrument, the hairs of the organ of Corti waited, like piano strings, to be "set into sympathetic vibration" (*Science and Culture* 60). Together, he proposed, these tiny hairs were accomplishing what a Fourier analysis could accomplish in mathematics. As each hair responded to its own natural resonance frequency, the organ resolved a complicated wave pattern into a number of simple, individual waves (*Science and Culture* 62–66). Like musical instruments, the human sensory organs and the nerves attached to them could respond only to vibrations of a particular type and only in a limited way. The signals they transmitted to the brain did not in the least resemble the stimuli that had excited them.

What, then, was the relationship between a nerve impulse and the external event that it represented? Like Müller before him, Helmholtz called activity in a nerve a *Zeichen* (sign or symbol) of its exciting cause.[18] "The sensations of our nerves of sense are mere symbols indicating certain external objects," he wrote, and it took a good deal of practice to learn how to interpret them (*Science and Culture* 66). In the eye, he pointed out, "the sensations of the optic nerve are for us the ordinary sensible sign of the presence of light," but other types of stimulation—pressure on the eyeball or an electrical shock applied to the optic nerve—could produce the same sign (*Sci-*

ence and Culture 152). It was up to the mind to interpret patterns of signs and to infer what they actually represented.

A central weakness of previous studies of perception, Helmholtz argued, was their failure to distinguish between the concepts of "sign" and "image":

> Insofar as the quality of our sensation gives us information about the peculiarity of the external influence stimulating it, it can pass for a sign—but not for an image. For one requires from an image some sort of similarity with the object imaged. . . . A sign, however, need not have any type of similarity with what it is a sign for. (*Science and Culture* 347)[19]

A picture, Helmholtz explained, was an "image or representation of the original" because it looked like what it was; it was "of the same kind as that which [was] represented" (*Science and Culture* 166). A nerve impulse, on the other hand, looked pretty much the same in every nerve, for every type of stimulus, and bore no mimetic relationship to the event that had inspired it. During the second—and successful—attempt to lay a transatlantic cable in the 1860s, physicist William Thomson (1824–1907) repeated a common eighteenth-century galvanic experiment by applying the telegraph wires to his own tongue. Thomson found that he could "taste" the differences among signals, and Helmholtz concluded that "nerve fibers and telegraph wires are equally striking examples to illustrate the doctrine that the same causes may, under different conditions, produce different results" (*Science and Culture* 150). What was language to a telegraph key was taste to the tongue. Like electrical fluctuations, nerve impulses produced the results that they did not because of what had excited them or even because of what they were but because of the device that was "reading" them.

As signs arbitrarily attached to the things they represented, nerve impulses resembled words in a language. It was no accident, submitted Helmholtz, that Thomas Young (1773–1829), who devised the first workable model for color vision, had also excelled at interpreting Egyptian hieroglyphics (Brazier, *Neurophysiology in the Nineteenth Century* 7; Helmholtz, *Science and Culture* 161). "There is a most striking analogy," declared Helmholtz, "between the entire range of processes which we have been discussing, and another System of Signs, which is not given by nature but arbitrarily chosen, and which must undoubtedly be learned before it is understood. I mean the words of our mother tongue" (*Science and Culture* 201). Justifying his comparison, Helmholtz pointed out that one had to learn the relationships between names and objects just as one had to learn the relationships between sensations and objects. "The words are arbitrarily or accidentally chosen signs," he asserted. "Each different language has different signs. Its understanding is not inherited" (*Science and Culture* 354–55). Once one established links between sensations and objects, these quickly became "just as firm and indestructible" as the associations between words and objects (Helmholtz, *Science and Culture* 201). Like nerve impulses, the sensations they made possible could be taken as "signs" of external reality.

Looking back on his studies of perception, Helmholtz reflected that "the impressions of the senses are only signs for the constitution of the external world, the interpretation of which must be learned by experience" (*Science and Culture* 390). Color, for instance, was a creation of the mind, a "sign" of physical properties that had nothing to do with the subjective sensation (Helmholtz, *Science and Culture* 168). Just as eighteenth-century writers compared language to clothes in which the truth might be "dressed up," Helmholtz envisioned sensations as the clothing of reality. In "The Facts in Perception" (1878), he wrote that objects in space "seem to us 'clothed' with the qualities of our sensations . . . although these qualities of sensation belong to our nervous system alone and do not at all reach beyond into external space" (*Science and Culture* 352). If all knowledge could be traced to one's sensations, and if these sensations were mere signs of external reality, then one could know only in a relative, never an absolute, sense. One could function only if one presumed that under the same conditions, for the same object, sensations would appear the same. What was so extraordinary was that given "so inconstant a system of signs," one could recognize objects and function in the world at all (Helmholtz, *Science and Culture* 173).

There was a significant difference, however, between Helmholtz's and his teacher Johannes Müller's uses of the term "sign." For both, it described the complex relationships among the external stimulus, the nerve impulse that transmitted it, the sensation registered by the sensory organ, and the mind's representation of that sensation. To Müller, the representation was to the sensation simply as "the sign for a thing" (Lenoir, "Eye as Mathematician" 117). Helmholtz, on the other hand, applied Hermann Lotze's idea of "local signs" to sensory physiology, arguing that people learned to associate signs with objects in the real world by studying how their eye and hand movements affected their positions relative to those objects (Lenoir "Eye as Mathematician" 122). In Müller's view, one could learn to read the nerves' signs but could never know the world in itself. Helmholtz, however, believed that one could use the relationships among signs to experiment on the world and gain meaningful knowledge of it.

Conclusion: The Language of Communication

Between 1850 and 1880, as Helmholtz's studies of visual and auditory perception became known throughout Europe, the question of what one could know about the world by interpreting its signs interested philosophers as deeply as it did physiologists. In "On Truth and Lies in the Extramoral Sense" (1870–73), Friedrich Nietzsche (1844–1900) proposed that all of people's "knowledge" of nature was based upon metaphor:

> The "thing in itself" (and that would be pure, inconsequential truth) is incomprehensible and utterly unworthy of effort for the creator of language as well. He designates only the relations of things to men and for their expression makes use of the most daring metaphors. First of all a nervous impulse is translated into an image. First metaphor. The image is again further formed into a sound! Second metaphor. (Kittler 187)[20]

Nietzsche, who had a profound interest in physiology and was well acquainted with Helmholtz's ideas about perception, argued that two representational systems mediated our access to reality: a sensory image, an arbitrary sign for the stimulus that had provoked it; and a word, an arbitrary sign for the sensory image.[21] To convey the degree of knowledge that this metaphorical system allowed, Nietzsche introduced his own metaphor: the mental image was to the "underlying neural activity" what Chladni's sound figures were to the actual tones they represented.[22] In identifying knowledge as metaphorical, he came remarkably close to Helmholtz, who regarded the nerve impulse and the sensation perceived as arbitrary signs of the original stimulus, analogous to words in a language.

Ironically, Nietzsche rejected Helmholtz's theory of perception—as he understood it—because he found it too restrictive. Helmholtz argued that we learn to perceive the world by making associations, by detecting recurring tendencies in patterns of signs. Because these physiological signs referred to the "untenable domain of things in themselves," he considered these associations to be "unconscious inferences" (Fullinwider 50, 47). Nietzsche inferred that Helmholtz saw perception as an unconscious process in which the signs one perceived could never reveal the actual nature of reality. Unable to accept this view, Nietzsche presented perception as a lively interaction of sensation and memory (Schlechta and Anders 108). As he saw it, "underlying neural activity" created a sensory image, and because the same neural activity always gave rise to the same image, there was a reliable correspondence between the activity and the image it produced—but this, of course, was Helmholtz's position as well.

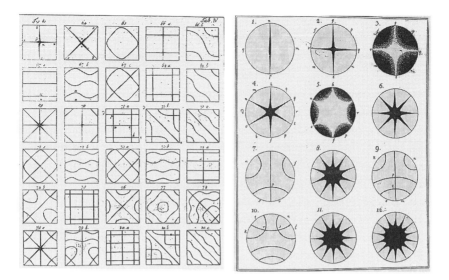

Ernst Florens Friedrich von Chladni's visual representations of sound, based on his analyses of sound waves. What these images were to sound, proposed Friedrich Nietzsche, mental images were to neural activity. (From Ernst Florens Friedrich von Chladni, *Theorie des Klanges* [Leipzig: Weidmanns, 1787] and *Die Akustik* [Leipzig: Breitkopf and Hartel, 1802].)

Nietzsche argues that it makes little sense to talk about "pure, inconsequential truth" when all knowledge relies upon metaphor, but he never denies that knowledge exists. The metaphorical scheme only precludes "knowledge" and "truth" as they have traditionally been understood: as objective, unmediated descriptions of reality. Helmholtz, a rigorous experimentalist, comes very close to Nietzsche in his assessment of what we can know. Both the scientist and the "philosopher with a hammer" see knowledge as metaphorical, mediated by systems of arbitrary signs, but both believe that knowledge is thriving anyway, fed continually by the systems of signs that make it possible.[23]

The real reason metaphors of writing and inscription pervade scientific discourse, Timothy Lenoir proposes, is that these references are "more than metaphoric" (*Inscribing Science* 1). They are not simplified translations used to communicate complex or abstract ideas to the public, nor are they decorative rhetorical figures added to engage readers. As can be seen by studying comparisons between organic and technological communications systems, metaphors do not "express" scientists' ideas; they *are* the ideas. Metaphors suggest new visions, images, and models; they inspire scientists to approach problems in new ways. To physiologists, the telegraph and associated studies in electromagnetism suggested the mechanisms by which the body transmitted information. To engineers designing telegraph networks, organic structures suggested ways to arrange centralized systems. More significantly, they motivated societies to establish more connections in hope of achieving a near-organic unity. Whether scientists thought visually, as Helmholtz and Morse claimed they did, or whether they thought verbally, comparisons between organic and technological systems underlay their thoughts about communication. In the nineteenth century, the real "language of communication" was metaphor itself.

Chapter 2
The Metaphoric Web

In the same lecture in which he called the nervous system a model for the telegraph, Emil DuBois-Reymond offered another, more elaborate analogy between organic and technological communications networks:

> Now, do you see the soul in the brain as the only sensitive, conscious region of the body, and the whole rest of the body as an inanimate machine in its hand? Just so the life of the great nation of France, otherwise centralized to the point of desolation, pulses only in Paris. But France is not the right analog; France is still waiting for a Werner Siemens to cover [überspinnen] it with a telegraph net. For just as the central station of the electric telegraph in the Post Office in Königsstrasse is in communication with the outermost borders of the monarchy through its gigantic web of copper wire, just so the soul in its office, the brain, endlessly receives dispatches from the outermost limits of its empire through its telegraph wires, the nerves, and sends out its orders in all directions to its civil servants, the muscles.[1]

Germany's government and social structure were superior to France's, DuBois-Reymond suggested, because the German communications system more closely replicated the body's own means of transmitting information. A telegraph network modeled on an organic system would allow a society to survive—and conquer—just as a sophisticated nervous system allowed a living animal to succeed.

As a metaphor for a communications system, the web is open to multiple interpretations and uses. On the one hand, it can represent the terrible efficiency of a power structure that commands its domain from a central point. On the other, it can be a liberating device through which scattered individuals can form associations and organize themselves as they see fit.

As a prominent scientist in a unifying nation, DuBois-Reymond recognized the benefits of centralized power, and he envisioned a communications net as an integral part of it. Whether or not this simplistic comparison reflects his actual view of the nervous system, he believed that it would appeal to his audience, who sought evidence that centralized power was a "natural" solution to organic as well as technical and political problems.

Bodies and nations need to transmit and coordinate information, and a web of connections seemed well suited for both activities. A society, however, is not a body, and DuBois-Reymond's analogy is an interpretation of organic communication that has its origin in technology and politics. The brain's "orders" can equally be viewed as responses; even as the brain influences, it is influenced in turn.

In the 1870s and 1880s, neuroanatomists thought carefully about whether the web was an appropriate model for the nervous system. Any image that represented the body's communications system had to account not just for its extraordinarily complex patterns of connections but for its abilities to transmit and store information. Could a web learn, remember, and think? When they scrutinized neural tissue, some investigators saw networks of interconnected cell processes, while others saw the tangled extensions of independent cells. Late nineteenth-century debates about a putative nerve network were not just scientific but philosophical and political arguments, inseparable from debates about personal and national identity. To call the nervous system a network is to read it, to interpret it, and interpretation varies greatly from culture to culture and individual to individual. Those who saw networks as unifying, empowering structures gladly imposed them on the nerves, but those who saw them as restrictive grids felt that they did nerve cells an injustice. Scientists agreed that the body's communications system worked because of its innumerable connections, but what did it mean to be "connected"? Did the cross-links of a network create or eliminate individual identity?

Eighteenth-Century Spiders

In his fascinating dialogue *Le Rêve de D'Alembert* (1769), Denis Diderot (1713–84) compares the human nervous system to a spider in its web. The dialogue takes place when Dr. Bordeu comes to examine the sleeping D'Alembert, who has passed the night in delirium and is still muttering intriguing philosophical thoughts. Mademoiselle de l'Espinasse, who has been caring for her partner, D'Alembert, and listening to his ravings throughout the night, becomes his interpreter as the materialistic doctor begins to challenge D'Alembert's ideas.[2] While the dialogue reflects the philosopher's delirium, careening from one subject to another, it remains focused on the problem of individual identity. The nervous system becomes a nexus through which the conversation repeatedly passes because for the materialist Diderot, the nerves unify the body. If there is such a thing as an individual—a possibility that the dialogue questions—it exists because of the way the nerves are organized. If one can accurately represent the nervous system, one can represent subjectivity itself.

In his unfinished physiology textbook, Diderot declares that the nervous system reminds him of a crayfish gathering information though its many feet (*Eléments de Physiologie* 87).[3] By bringing information to a central coordinating point, it makes the body into an organic whole. In *Le Rêve de D'Alembert,* Diderot's metaphors for the nerves are equally compelling, always stressing how the body controls elements that might otherwise remain independent. Mademoiselle de l'Espinasse offers the two central comparisons, at first suggesting that an individual is like a swarm of bees clustered around a branch like a bunch of grapes. Each "grape" is "a being, an individual, some kind of animal," and each resembles all the others, but if one bee stings another, the entire swarm suddenly becomes agitated as if it were a single living animal (45–46). Dr. Bordeu replies that this is a mistaken impression: a unified animal differs markedly from a swarm of independent living particles, for its elements are continuous, not just contiguous (47).

In real animals, the characters decide, a web of nerves unifies the organism by turning contiguity into continuity. At this point, a new metaphor emerges: Mademoiselle de l'Espinasse compares the mind and body to a spider in its web. The fact that D'Alembert regains consciousness and interrupts the speakers just as she is voicing this idea shows its close relation to the central issue of identity. The spider's web, in which information about a given area is transmitted to a central point, suggests not just how the body is organized but how subjectivity is established. This is the topic that the characters most want to explore. "It is the constant, invariable relation [*rapport*] of all impressions to this common origin that constitutes the unity of the animal," proclaims Bordeu (94).

Diderot's characters depict the web of nerves as an instructional device and an alarm system, always stressing its role in delivering information. In such a system, the subject appears to be the spider, the one who sits in the center, who is informed, and out of whose body the threads grow. Anticipating Proust, Mademoiselle de l'Espinasse confesses that when she lies alone in the darkness on the verge of sleep, "I exist as if on a point; I almost cease to be material, I feel only my own thought" (Diderot 98; Fellows 104). She also suggests that identity is based on a person's memory of her past experiences (94). Mademoiselle de l'Espinasse's insights about her mind and body imply that the "origin of the network," which Dr. Bordeu presents as the site where sensory impressions are developed into thought and reason, is not sufficient to create a human subject (94). Individual identity involves not just the processing center but the web as a whole.

Although Diderot's characters emphasize the web's informative aspects, Dr. Bordeu does briefly refer to its role as a disseminator of commands. An animal, the characters decide, must live in a state of either despotism or anarchy, and the web ensures that order will be maintained. "The origin of the bundle commands, and all the rest obeys," proclaims the doctor. "The animal is master of itself" (115). In his physiology textbook, Diderot is even more explicit, calling nerves the "ministers," the "servants," and even the "slaves" of the brain (*Eléments de Physiologie* 88, 90). If not properly controlled by a central governor, these servants can become despots. The body remains healthy only as long as the brain maintains command; illness, particularly nervous illness, is a kind of anarchy.

Interestingly, Diderot envisions a system in which each sensory nerve is connected to the brain but the conducting threads are never connected to each other. Differing sharply from his nineteenth-century successors, he declares that "if there were an anastomosis [a physical connection] between the nerves, there would no longer be any order [*regle*] in the brain, the animal would be mad" (*Eléments de Physiologie* 92). If nerves carrying information could communicate before their messages reached the command center, he believed, central control would break down. The structure that Diderot describes in his physiology text is thus not a true network, for it contains no cross-links, only direct lines between the center and the periphery.

When one regards the web from this perspective (which Diderot, in his philosophical writing, resists doing), subjectivity becomes much less complex, much more limited. It is collapsed from a sensitive network to a single point issuing orders. Taken as a whole, *Le Rêve de D'Alembert* challenges the simplistic view that the individual is the central point dispatching commands and receiving information. Diderot's representation of the nervous system as a spider in its web stresses the sensitivity and superbly adapted structure of the nerves, their ability to unite the organism by providing information from all of its parts.

One year after Diderot wrote *Le Rêve de D'Alembert*, Paul D'Holbach (1723–89) made use of its central metaphor in his *Système de la Nature* (1770):

> In man, the nerves come to reunite and to lose themselves in the brain. This organ is the true location of feeling; just like the spider that we see suspended at the center of its web, it is promptly notified of all the marked changes

which occur all over the body, even in the extremities to which it sends its threads or branches.[4]

Like Diderot, D'Holbach emphasized the sensory aspects of the body's web, its role as a protective alarm system to alert the animal of potentially dangerous movements. As nineteenth-century writers would find, the spider's web was the ideal metaphorical vehicle for the nervous system not just because of their common function, the ability to transmit information to a central point. The two had a common structure as well, occupying vast spaces with tiny, interconnected threads. For nineteenth-century neuroanatomists struggling to map out the nervous system's connections, the web and the network became irresistible images.

The Networks of Physics

While physiologists like Helmholtz and DuBois-Reymond studied the functions of nerves and muscles, anatomists were performing closely related studies of their structure. What did the body's networks look like? How were its nerve centers linked, and how were its nervous pathways organized?

Like the question of how individual nerve cells transmitted signals, the problem of how these cells were physically arranged was formulated in terms of electricity and magnetism. In particular, the question of connections seemed inseparable from the issue of how forces acted over a distance. Since the time of Newton, physicists had doubted whether one object could exert influence over another through empty space. Newton himself had rejected the possibility of action across a void, proposing that space as we know it is filled with "ether," a rarefied material of infinitely fine particles. Like the physicists, nineteenth-century neuroanatomists wondered what exactly it meant to be in contact, to communicate a force to another body. Even in Galvani's day, Alexander von Humboldt had hypothesized that nerve fibers were surrounded by "sensible atmospheres" and could exert a "galvanic influence" on one another even though they were not in physical contact (Müller 1: 683). To transmit information, did nerve fibers have to touch? And if they did not, how did their messages move across the intervening space?

In the 1840s, physicist Michael Faraday rejected the idea of action at a distance, maintaining that a model featuring isolated, individual atoms separated by empty space could not account for electrical and magnetic laws. If space and matter were two distinct entities, he reasoned, space would "permeate all masses of matter in every direction like a net, except that in place of meshes it w[ould] form cells, isolating each atom from its neighbors, and itself only being continuous" (Gillispie, *Edge of Objectivity* 454). As Faraday's comparison shows, a net can represent not just a connecting structure but a grid that locks individuals into position and separates them from one another. Written in 1844, his description suggests the cellular structure of early nineteenth-century institutions (prisons, schools, hospitals, and factories) designed to control individuals by assigning them fixed positions in space and limiting their communications (Foucault 143).[5] For Faraday, the network image brought to mind this sort of isolating grid, a model that in the realm of physics did not work.

When Faraday struggled to envision what a magnetic field actually involved, however, he described it as an organic web, a space full of lines and "tubes of force." For Faraday, these lines were very real, because without them he could not account for the influence of one object on another that created electrical and magnetic induction (Gillispie, *Edge of Objectivity* 452–55; Brody, "Physics in *Middlemarch*" 46). Faraday's "lines of force," an attempt to show how magnets exerted their forces at the microscopic level, became one of the most influential models of the nineteenth century.

As writers from many fields struggled to describe how little-known forces exerted their influence, they fell back on Faraday's image of tiny, hairlike lines. Herbert Spencer's (1820–1903) description of how the nervous system creates its pathways, for instance, reflects both Faraday's electromagnetic models and engineers' diagrams of the railways for which Spencer once worked: "lines of nervous communication will arise . . . and will become lines of more and more easy communication in proportion to the number and strengths of the discharges propagated through them" (Shuttleworth 163). Whether the communications system is organic or inorganic, Spencer implies, patterns of use will affect the emerging structure.

Other writers, both scientists and nonscientists, drew a parallel between magnetic fields and the influences (popularly known as magnetic forces) exerted by powerful minds. British physicist John Tyndall (1820–93), a good popularizer as well as a gifted experimentalist, explained that when a piece of steel was magnetized, "nothing [was] actually transferred" from the magnet to the steel; instead, the magnet was rearranging matter that was already present. The magnet, he declared, did not lose any more force than "I should lose, had my words such a magnetic influence on your minds" (1: 355). Considering the degree to which innate and cultural forces could determine individual actions, Walter Pater (1839–94) wrote that "[necessity] is . . . a magic web woven through and through us, like that magnetic system of which modern science speaks, penetrating us with a network, subtler than our subtlest nerves, yet bearing in it the central forces of the world" (Pater, "The Renaissance," 682). While magnetism meant different things to physicists, mesmerists, and the public at large, it proved an ideal metaphor for discussing influence of any kind, whether exerted by a coil of wire, a human mind, a society—or a neuron.

The Networks of the Nerves

In nineteenth-century neuroanatomy, the most highly contested issue was how nerve cells were connected to one another. Debates about whether neurons were independent cells or elements in a physically continuous net reached their peak in the late 1880s, but the problem of how signals moved from one element to the next had occupied scientists as early as 1839. While discussing the functions of posterior and anterior roots in the spinal cord, Claude Bernard specified, "all we are really talking about here is connections" (1: 23). A central issue of nineteenth-century physics as well as of neuroanatomy, "connections" challenged assumptions about spatial separation and individual identity.

While increasingly sensitive microscopes and staining techniques allowed scientists to observe nerve cells in greater and greater detail, improvements in technology never led immediately to any widely accepted new theories of structure and function. Instead, new modes of observation introduced new controversies and new questions. In the most famous case, Italian neuroanatomist Camillo Golgi (1843–1926), who invented a special silvernitrate staining technique, observed that nerve cells merged to form a continuous net. Spanish neurohistologist Santiago Ramón y Cajal (1852–1934), who relied on Golgi's stain, saw independent cells with freely terminating branches.

Both those who argued for autonomous nerve cells (neuronists) and those who argued for a nerve net (reticularists) accused their opponents of "physiological prejudices." Rather than looking in a careful, unbiased way at the structures present, they claimed, the others were seeing imaginary structures that confirmed their theories about the way the nervous system functioned. Golgi accused the neuronists of physiological bias, and Cajal accused the reticularists of failing to free themselves from prevailing ideas, even of succumbing to hypnotic suggestion. These charges raise a vital question: what exactly determines what scientists see under their microscopes?

Such marked dissent among neuroanatomists using similar techniques suggests that their perceptions were determined by more than their lenses and stains. What would make a nineteenth-century neurohistologist *want* to see a net rather than a collection of independent cells?

In 1838, Robert Remak (1815–65) changed scientists' vision of neural "circuitry." He demonstrated that cell bodies—clustered together in ganglia—and axons—elongated processes for transmitting impulses over distances—were actually parts of the same cells and "were in direct communication" (Clarke and O'Malley 46; Brazier, *Neurophysiology in the Nineteenth Century* 136). Theodor Schwann's (1810–82) proposal the following year that all animal tissues were composed of individual cells encouraged scientists to view the cell, not the fiber, as the nervous system's essential unit. In 1839, Johannes Müller wrote, "we know, from microscopic examination of the nerves, that the primitive fibers do not unite with each other, but remain distinct from their origin to their termination" (1: 773). Remak, Schwann, and DuBois-Reymond all studied with Müller in Berlin, and it has been suggested that DuBois-Reymond could imagine the possibilities of animal electricity to a much greater extent than Galvani because the new notion of individual nerve cells freed him from envisioning each fiber as a conduit directly connected to the brain (Brazier, *Neurophysiology in the Nineteenth Century* 77).

Neurons differed from other cells in the great variety of processes that emerged from their rounded cell bodies, and in 1865 Karl Deiters's (1834–63) studies showed that these processes were of two distinct types: a single, elongated "axis cylinder" (axon); and a much finer, more highly branched collection of "protoplasmic processes" (dendrites) (Brazier, *Neurophysiology in the Nineteenth Century* 138). Joseph von Gerlach (1820–96), who introduced anatomists to ammonium carminate and goldchloride staining, was able to observe the ways in which the two types of processes terminated. In 1872, he declared that nerve cells were physically joined, the finest branches of their protoplasmic processes merging to form a continuous network. Because the axis cylinders also apparently merged to form another, coarser net, the large fibers Gerlach observed in the brain seemed to have a double origin, in a coarsely woven network at one end and in an "extremely fine-meshed network" at the other (Clarke and O'Malley 88). Ultimately, every element of the nervous system was physically connected to every other, so that it constituted a vast communications net. From this point on, the idea of a continuous net of nerve fibers gained more and more adherents, and for almost two decades it supplanted the image of the individual nerve cell.

While the nerve-net paradigm predominated in the laboratories of the 1870s and 1880s, it never fully suppressed the belief in autonomous, freely terminating nerve cells. By the late 1880s, many scientists independently challenged the reticular hypothesis, and the reticularists answered their attacks. Although many prominent neuroanatomists took part in these debates, I will focus on the arguments of Camillo Golgi, "the most ardent and influential proponent of the nerve net," and Santiago Ramón y Cajal, its most aggressive opponent (Clarke and O'Malley 90). Both Golgi and Ramón y Cajal were superb anatomists; both used the same basic technique to stain their preparations; and both believed that they were viewing cells exactly as they were, having freed themselves from all physiological and cultural prejudices. Even when they shared the Nobel Prize for medicine in 1906, however, they continued to challenge each other's observations.

In 1873, Camillo Golgi's new silver stain made it possible to see all of a cell's processes with perfect clarity. When a thin section of neural tissue is treated with silver nitrate, Golgi discovered, then soaked in a mixture of Müller's solution and osmic acid, a few individual cells will absorb the silver, so that it forms deposits throughout all of their branches. This technique allows one to see their shape in perfect detail just as one can see every twig of a tree in winter (Ramón y Cajal, "Conexión general" 24). As Michael Hagner has shown, however, Golgi's 1873 article describing his stain initially received no response. Written in Italian, the essay contained no illustrations demonstrating the technique's potential and failed to stand out from numerous other articles of the 1870s announcing promising new stains. Possibly Golgi himself did not want the technique publicized. Still used by neuroscientists today, the Golgi stain is an invaluable tool for the same reason it first seemed unreliable: due to unknown causes, only some cells absorb it (Hagner, "Les Frères ennemis," 40).

A cautious scientist, Golgi remained modest about his ability to resolve neuronal structures. As he expresses it in his publications, his real goal was to study the way neuronal processes "behave" (*se comportent*), and he follows them to their terminations with what he believes to be a fully open mind (Golgi, "Recherches sur l'histologie" 305; Golgi, "La doctrine du neurone" 13). Just because one cannot see the ends of the ramifications, he warns, this does not necessarily mean that they merge into a net. Describing cells in the cerebral cortex, he writes that "it is probable that these innumerable subdivisions anastomose among one another and thus form a genuine plexus, but the extreme complexity of this network does not permit me to make any positive statement on the subject" ("Recherches sur l'histologie" 297–98; Clarke and O'Malley 93).[6] Even in 1906, when the idea of independent nerve cells had gained general acceptance, Golgi claimed that the "so-called independence" of neurons was based on nothing but "the impossibility of verifying the most intimate connections" ("La doctrine du neurone" 7). Based on his own, careful observations of the brain, Golgi believed that those who saw autonomous neurons were not observing closely enough.

As soon as Golgi began using the silver stain, he rejected Gerlach's claim of a dendritic net. Expressing grave doubts about Gerlach's gold-chloride preparations, Golgi suggested that Gerlach *believed* he saw a net made of fused protoplasmic processes because spinal reflexes demanded it. Gerlach was proposing a structure based on physiological need, not anatomical fact. If one restricted one's observations to the appearance and "behavior" of dendrites, Golgi maintained, it was not clear that they were involved in communicating signals at all. Instead, their frequent association with blood vessels suggested a nutritive function ("Recherches sur l'histologie" 295; "La doctrine du neurone" 19; Clarke and O'Malley 92).

Despite his insistence on empiricism, Golgi thought constantly about physiology. As he studied neuronal processes, he wondered how they communicated with one another. If there were no dendritic net, how were nerve cells connected? Did communication require physical contact, or might signals somehow jump from cell to cell? In a review article of 1883, Golgi posed this question to himself and his fellow scientists:

> If nerve fibers proceed neither directly nor indirectly from the protoplasmic extensions, and if there is no communication between the different groups of cells of the nervous system, either by way of anastomoses or by the diffuse network, what then is the mode of origin of the nerve fiber in the gray matter? How then is a functional relationship, which one is forced to admit *a priori*, established between the various cells of different parts of the nervous system? ("Recherches sur l'histologie" 95; Clarke and O'Malley 92)[7]

By taking the "functional relationship" as a given, Golgi is committing the sin—if it is a sin—of which he accuses Gerlach. He is approaching anatomical structures with a preconceived notion of the activities they must perform.

While Golgi presupposes that neurons communicate, he remains open-minded about the *way* they communicate. He is never certain whether the transmission of signals requires actual, physical contact, and like the neuronists Ramón y Cajal and August Forel (1848–1931), he offers the example of electrical induction as evidence that forces can be transmitted between structures not physically connected. "Since at the moment electrical studies

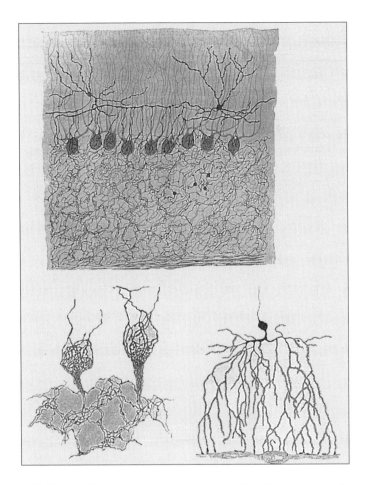

Individual cells, or a neural net? Camillo Golgi's drawings of cells in the cerebellum. *Above,* "Bundles of small fibers, coming from neural processes of small cells in the molecular layer, which supposedly form the so-called terminus on the surface of Purkinje cells, [whose large, shaded cell bodies lie under the darker fibrils] continue through an infinite number of subdivisions into a neural net." *Lower left,* a more detailed picture of "fibrils emanating from the molecular layer, spreading themselves out over the surface of Purkinje cells." *Lower right,* a cerebellar cell "feeding" on a blood vessel through its dendrites. (From Camillo Golgi, "La Doctrine du Neurone," *Les Prix Nobel 1906* [Stockholm: Imprimerie Royale, 1908], 4, 5, 19. My translation.)

are demonstrating that electrical currents can occur without direct continuity of the conducting parts," he argues in 1891, "why can we not admit that
the same laws are also valid for the nervous system?"[8] Although a nerve net
would seem to involve direct, physical continuity between neurons, Golgi
never makes it clear whether by "continuity" he means a structural or functional link, a material merging, or the ability to conduct signals reliably. In
his Nobel Prize acceptance speech, he claims that if one accepts the idea of
a complex neural net extending throughout the gray matter of the brain,

> It [is] no longer necessary to invoke a material connection, a fusion between
> one fiber and another, to account for the functional relationships that con-
> nect [qui courent entre] different groups of cells and between the different
> parts of the central nervous system. . . . there [is] no reason to believe that
> direct continuity between fibrils of different origin is the condition *sine qua
> non* for the transmission of excitation from one to another.[9]

Like Gerlach, Golgi is committed to the idea of a neural net not just because
of what he sees but because of the way he believes the brain works. To
Golgi, the notion of a complex, diffuse nerve network providing connections among all the elements in the brain eliminates the need for "hard
wiring."

While he argued vociferously against Gerlach's putative dendritic net,
Golgi did observe a coarsely woven axonal net composed of fused axis
cylinders. Golgi identified two types of neurons: Type I, which produces one
long, continuous axis cylinder; and Type II, "whose neuronal process sub-
divides itself in a complicated manner, loses its individuality, and takes part
in toto in the formation of a neuronal reticulum which traverses all the layers of the gray matter."[10] Based on their distribution, Golgi proposed that
Type I cells were motor neurons, carrying information outward from the
central nervous system, whereas Type II cells (those that "lost their individuality" by merging into a common network) were sensory, conveying information from the periphery to higher processing centers ("Recherches sur
l'histologie" 301; "La doctrine du neurone" 12). Many scientists took issue
with Golgi over this broad, speculative association of structures with functions, protesting that the two types of neurons were interspersed throughout
the nervous system. But that, argued Golgi, was exactly the point. The system worked so well because all of its cells were interconnected, so that no
particular function could ever be tied to a given locus.[11]

All elements of the nervous system, Golgi believed, contributed to the
neural network, and it was the *network*, not the individual cell, that carried
out the system's functions ("La rete nervosa" 592; "La doctrine du neu-

rone" 6). In the opening of his Nobel Prize acceptance speech, Golgi declared that he had always opposed the neuron doctrine (the idea that the nervous system consisted of independent cells). He particularly objected to the idea of the cell as "an elementary, independent organism" ("La doctrine du neurone" 2). As an anatomist, Golgi had observed both individual nerve cells and a nerve net, and his ideas about brain function played as strong a role as his laboratory experiences in his adoption of the reticular hypothesis. Extending throughout the gray matter of the central nervous system, the neural net provided essential "physiological communication between the sensitive and the motor spheres" ("Recherches sur l'histologie" 302). When properly understood, Golgi believed, the net should be regarded as an organ in itself, "an organ playing a fundamental role in the specific function of the nervous system" ("La doctrine du neurone" 14). Both Golgi's anatomical studies and his physiological thinking (which were, in the end, inseparable) convinced him that nerve cells worked collectively, not individually. He concluded his acceptance speech by asserting, "I cannot abandon the idea of a unitary action of the nervous system" ("La doctrine du neurone" 30).

If anatomical studies showed anything with certainty, Golgi argued, it was that the nervous system was designed to maximize the number of possible connections among cells. The concept of "isolated transmission," the conveyance of a signal directly and exclusively from one cell to another, had no basis in anatomical fact. Instead, any given nerve cell was connected to vast numbers of other cells, some of them in distant regions of the system. "In most neural centers," Golgi wrote, "what takes place is in no case an isolated and individual relationship between the fibers and the cells, but, on the contrary, an arrangement clearly meant to [destinée à] permit the greatest variety and the greatest complexity of their mutual relationships. . . . *The fundamental arrangement of the central elements indicates a tendency to carry out the most extensive and complex communications, not limited and isolated relationships.*"[12] Because all nerve cells were interconnected through a common structure, activity in any sensory receptor could affect an "infinite number" of neurons in the central nervous system, and a central cell could affect an infinite number of cells in the periphery (Golgi, "La rete nervosa" 594). With such an arrangement, the attempt to associate particular structures with functions simply made no sense. Like the idea that cells communicated with one another through exclusive, "isolated" links, brain localization ignored the extraordinary complexity of relations among neurons, denying reality in the interests of simplifying physiological theory. Both anatomy and physiology indicated that the nervous system worked as a whole.

Communication without a Net

In the late 1880s, a number of scientists independently challenged Golgi's nerve-net hypothesis on both anatomical and physiological grounds. Swiss embryologist Wilhelm His (1831–1904) observed the development of neuronal processes and declared in 1887 that "every nerve fiber arose from a single nerve cell" (Clarke and O'Malley 99). As a specialist in developmental neuroanatomy, His believed in action at a distance, writing that "the continuity between two tracts is not mandatory for the explanation of the influence of one fiber system upon another . . . within a certain area, the stimulus which is transmitted by a neighboring tract can be communicated to various adjacent pathways" (Clarke and O'Malley 101–2). When combined with His's anatomical observation that each fiber could be traced to one specific nerve cell, this physiological hypothesis that structures did not need to be in physical contact to influence one another suggested that a system of independent neurons could function as effectively as a network of conjoined cells.

During the same year that His was studying neuronal development, Swiss psychiatrist August Forel reached the same conclusions by observing neuronal degeneration. Using Golgi's silver stain, Forel showed that whenever he separated a neuronal process from the cell body to which it was attached, it would waste away, having lost its nutritional center. Whenever Forel cut an axon or good-sized dendrite that seemed to "lose itself" in a net, only some of the fibrils in the "net" would die—those attached to that particular cell. Forel concluded that "a nerve network does not exist, and each nerve cell is in contact with, but not in continuity with, its neighbor" (Clarke and O'Malley 104). In challenging Gerlach's and Golgi's hypothesis, Forel drew upon his knowledge of electricity as well as his neuroanatomical observations, writing that "electricity gives us such innumerable examples of similar transmissions without direct continuity that it could well be the same in the nervous system" (Clarke and O'Malley 105–6). Electromagnetic induction, which had helped DuBois-Reymond construct a physiological model for nerve-impulse transmission, now seemed to legitimize an anatomical model for neuronal connections. So appealing was the analogy that it was used by scientists both advocating and opposing the nerve net.

While His and Forel encouraged many scientists to rethink their positions on neuronal connections, the definitive evidence for cellular "independence" was still lacking. Scientists needed an anatomical preparation that clearly demonstrated freely terminating processes. Spanish neurohistologist Santiago Ramón y Cajal provided this evidence with his studies of the bird cerebellum in 1888 and 1889.[13] At the time he undertook his studies, Ramón y Cajal was not aware of His's and Forel's work, just as they had been unaware of each other's projects. Within two years, three scientists, working independently and approaching the problem of neuronal connections from different perspectives, rose to challenge the reigning hypothesis of the nerve net. In 1891, German anatomist Wilhelm Waldeyer (1836–1921) coined the term "neuron" and proposed that nerve cells were autonomous, bounded structures. After this, most scientists—except Golgi—gave up the idea of a neuronal reticulum. But considering that the anatomists whose work overthrew the reticular hypothesis used Golgi's staining techniques, what other factors allowed them to challenge an idea that they had embraced so strongly for almost two decades?

When Ramón y Cajal criticizes Golgi's continued advocacy of the reticular hypothesis, he sounds a great deal like Golgi himself. Believing that he is seeing neuronal connections exactly as they are, he aims to "free [*desligar*] histology from every physiological compromise" (Ramón y Cajal, "Conexión general" 480).[14] According to Ramón y Cajal, it is Golgi whose vision is clouded by physiological prejudice. As an example, he cites Golgi's facile identification of Type I cells as motor and Type II as sensory, warning that one can never identify a cell's function based on observations of its structure ("Conexión general" 486).

Ramón y Cajal regarded the nerve-net hypothesis as an insidious idea that kept scientists from seeing and thinking clearly, comparable to any other social belief that hinders productive action. In his publications, he calls the hypothesis "despotic" and "seductive" and even compares it to a hypnotic suggestion ("Conexión general" 21; "¿Neuronismo o reticularismo?" 215). Expressing pity for Golgi, whose techniques he admires, he describes him as a scientist who succumbed to the prejudices of his environment. "It is a sad truth," he writes, "that almost no one can separate himself [*sustraerse*] entirely from tradition and the spirit of his times. In spite of his great originality, Golgi suffered considerably from the suggestive power of Gerlach's diffuse, interstitial nets" ("¿Neuronismo o reticularismo?" 221). What Ramón y Cajal does not explain is how he extricated himself from this net of suggestion. Like Golgi, he believed that his opponents' views of structures were distorted by fixed ideas, whereas he was seeing neurons with perfect clarity.

Besides being prejudiced, according to Ramón y Cajal, scientists who thought they were seeing a net were just plain lazy. What they actually encountered when gazing through their lenses was an extraordinary tangle of extremely fine processes. Rather than struggling to resolve these processes, carefully following each to see to what it was attached, they opted for a much easier solution, declaring that the processes physically merged. In his seminal article on the bird cerebellum, Ramón y Cajal wrote that "the fibers are inter-woven [*se entrelazan*] in an extremely complex way, giving rise to an intricate and densely woven plexus, but never a net" ("Estructura de los centros nerviosos" 314). Just as the physiologist Helmholtz had invoked language systems when discussing the brain's functions, Ramón y Cajal compared himself to Champollion in his efforts to make sense of its structures: "Champollion, trying to decipher the dead language of Egyptian hieroglyphics, and Layard and Rawlinson, trying to pull [*desentrañando*] the mysterious meaning out of the cuneiform characters in the inscriptions of Nineveh and Babylonia, have posed problems much simpler than neurologists have."[15] Like a language, the nervous system worked through intricate and complex relations. Ramón y Cajal agreed with Charles Babbage that any code could be cracked in time, but one could only discover the patterns in these relations if one was willing to invest the effort.

Ramón y Cajal thus argued for cellular autonomy for many of the same reasons that Golgi advocated a net. His opponents were failing to see structures as they were because of physiological and cultural prejudices and because of their own lack of motivation. If one could see the nervous system clearly, Ramón y Cajal insisted, one would observe that "every element is an absolutely autonomous physiological canton" and that every cell "always maintains [*conserva*] its individuality" ("Estructura de los centros nerviosos" 314; "Conexión general" 479–80). In his highly regarded histology text-book, Ramón y Cajal traced the development of cell theory from its origins in the 1820s, stressing its importance for contemporaneous neuroscience. In the late 1830s, German anatomist Theodor Schwann had described cells as "live individuals associated among themselves" (Ramón y Cajal, *Elementos de histología normal* 148). Ramón y Cajal himself depicted them as distinct, clearly defined entities that are somewhat, but never fully, independent of the larger organism in which they live: "living beings that, in the midst of their subordination to an organic whole, enjoy a certain functional autonomy" (*Elementos de histología normal* 189). Because of their membranes, cells were always "correctly separated from one another," and the idea of an independent identity for each cell rested largely on the fact that these membranes remained intact and their contents never physically merged (*Elementos de histología normal* 156). Golgi shared this notion of independence, for he invariably described cells that merge into a net as "losing their individuality." Ramón y Cajal, however, never saw neurons merge. Viewing the cell as "an organism in miniature," he pointed out that each one could manifest its

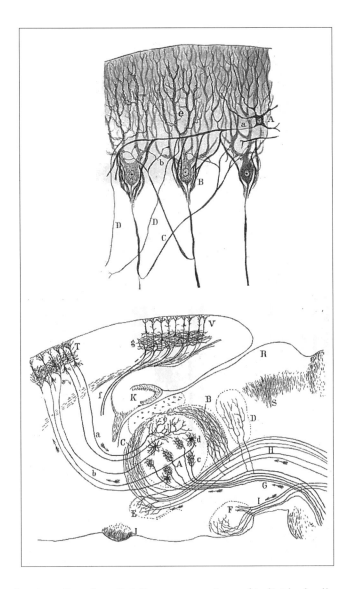

Santiago Ramón y Cajal's representations of individual cells
in the cerebellum. *Above,* the highly branched dendritic
"trees" of Purkinje cells (*A*) and the axons of cells in the
plexiform layer whose descending collateral branches form
"baskets" around the Purkinje cell bodies. *Below,* a diagram
of neuronal connections in a rabbit's sensory thalamus.
(From Santiago Ramón y Cajal, "Structure et Connexions
des Neurones," *Les Prix Nobel 1906* [Stockholm:
Imprimerie Royale, 1908], plates 1, 13).

own, distinctive behavior (*Elementos de histología normal* 155). One of the strongest pieces of evidence for the neuron doctrine was the physiological fact that not all cells responded identically to the same stimulus (*Elementos de histología normal* 190).

While Ramón y Cajal disagreed with Golgi about the ways nerve cells were connected, he agreed completely about the complexity of neuronal connections. Like Golgi, he believed that each nerve cell was in contact with countless others. In a speculative article of 1895 in which he discussed how the brain formed associations, Ramón y Cajal proposed that the cells of sensory organs activated the neurons of higher centers in an "avalanche" pattern, each individual sensory neuron activating vast numbers of cells in the brain ("Algunas conjeturas" 3). Ramón y Cajal supported Golgi's view that most connections in the nervous system were not individual but collective. In carrying out neural functions, vast groups of cells acted in concert. He differed with Golgi only in his thinking about what kind of connections would best account for the behavior of the system. For Golgi, only a network could carry out such complex, collective actions. Ramón y Cajal, however, was convinced that they could be accomplished with "living individuals associated among themselves."

Like His, Forel, and Golgi himself, Ramón y Cajal thought a great deal about what a neuronal "connection" involved. Were nerve cells in "contact" if they were contiguous, their processes terminating extremely close to those of adjacent cells, or did "contact" mean that they must physically touch? In an article written in the last year of his life, Ramón y Cajal voiced the problem as follows: "Do [the final neuronal arborizations] really touch the naked protoplasm of the cell, or do delimiting membranes exist between both elements of the synapse?"[16] This question went to the very heart of the way that the nervous system worked.

In his discussions of neuronal connections, Ramón y Cajal usually described associations in terms of "very intimate contact" ("Croonian Lecture" 453). He carefully distinguished such contact from anastomosis, the physical joining of very highly branched structures at their finest points. Ramón y Cajal agreed with Golgi and Forel that the laws of electricity proved it was not necessary for two elements to be physically connected in order to influence one another. In his important study of the bird cerebellum in 1888, he proposed that he was failing to observe physical links between processes that "lost their individuality" either because Golgi's silver stain did not reveal them or because "the connection between them and the axis cylinders can be mediated and the neural action can occur just as the electrical currents in inducting cables act on wires in which currents are induced."[17] Physics suggested that communication did not require physical continuity.

Like other anatomists of his day, Ramón y Cajal thought constantly about how the cells he was observing might transmit impulses. Until the 1950s, neurophysiologists did not know how neurons conducted their impulses or how signals were transmitted from cell to cell, but by drawing on his knowledge of chemistry and physics, Ramón y Cajal was able to infer what must be involved. In his neurohistology textbook of 1899, he speculated that:

> the very fact of the transmission of the wave from one neuron to another would obey the laws of chemical phenomena: in reality, the impulse provokes a chemical change in the neuronal arborizations which, working in turn as a physico-chemical stimulus on the protoplasm of other neurons, would create new currents in these. The state of consciousness would be precisely tied to these chemical changes brought about in the neurons by these nerve endings.[18]

This account of synaptic transmission requiring no physical contact would still be acceptable to most neuroscientists today. As chemical and physical experiments suggested, one body could influence another without ever touching it, and it seemed reasonable to presume that living cells followed the same laws.

Like Golgi, Ramón y Cajal pointed out that the nervous system was designed to maximize such influence, the elaborately branched dendritic trees of neurons offering the greatest possible number of contacts with other cells. The more extensive a neuron's ramifications were, the greater was its ability to influence other neurons ("Croonian Lecture" 465). Both Golgi and Ramón y Cajal admired the nervous system's rich, complex patterns of connections, but while Golgi used them to argue for a net, the Spanish scientist used the same facts to argue for cellular independence. Ramón y Cajal emphasized a quality of the nervous system that Golgi never stressed: its extraordinary plasticity.

In histology, he warned, "it is impossible to separate the static from the dynamic . . . form is an unstable property" (*Textura del sistema nervioso* viii). Fundamental to the nervous system was its ability to learn and forget, and even if one ignored this fact as a "physiological prejudice," it was clear that neurons themselves constantly grew and changed. A strong believer in willpower and individual initiative, Ramón y Cajal felt that vigorous mental exercise improved one's ability to think by creating new connections among dendritic trees. In his Croonian lecture of 1894, he even went so far as to speculate that intensive thought encouraged the growth of new collaterals (466). For Ramón y Cajal, the nervous system was a living, dynamic structure that could grow and change so dramatically because it was composed of independent, freely terminating elements.

Like Helmholtz, Ramón y Cajal compared the nervous system to a telegraph network, not only in its function but in its structure. The enormous Purkinje cells of the cerebellum, he remarked, supported the fibers of granular cells "something like the way that telegraph posts support the conducting wire" ("Conexión general" 484). When one considered all the different functions of a neuron, he suggested in his Croonian lecture, it became clear that the neuron was a sophisticated communications device. "The nerve cell provides an apparatus for the *reception* of currents," he wrote, "manifested by the dendritic expansions and cell body; an apparatus for *transmission,* represented by the prolonged axis cylinder; and an apparatus for *apportionment* or *distribution,* represented by the arborization of the nerve terminal."[19] To Ramón y Cajal, a neurohistologist, it seemed quite reasonable to compare the fundamental unit of an organic communications system to a single transmitter or receiver in a vast technological communications network.

In the conclusion of the same lecture, however, Ramón y Cajal rejected the telegraph metaphor, finding the technological network an unworthy analog for a living, dynamic communications system. Such a comparison, he found, inevitably brought to mind Golgi's net:

> In relation to the theory of networks, the theory of free arborizations of cellular expansions capable of growth appears not only more probable, but also more encouraging. A preestablished, continuous network—a sort of grid of telegraph wires in which one can create neither new stations nor new lines—is something rigid, immutable, incapable of being changed, that hurts the feeling we all have that the organ of thought is, within certain limits, malleable and capable of perfection, above all during the time of its development, by means of well-directed mental gymnastics. If we were not afraid of abusing comparisons, we would defend our point of view by saying that the cerebral cortex is like a garden filled with innumerable trees, the pyramidal cells, which, thanks to intelligent cultivation, can multiply their branches, drive in their roots more deeply, and produce ever more varied and exquisite flowers and fruits.[20]

In the early 1850s, the telegraph network had appealed to Helmholtz and DuBois-Reymond as an excellent metaphorical vehicle for the nervous system, for at this time the young, rapidly growing web had seemed almost alive. By 1894, with most lines and stations already in place, the same network struck Ramón y Cajal as rigid and dead. Perhaps the quality of technological communications networks that most appealed to physiologists was their dynamism, a characteristic that disappeared once they were fully constructed. To Ramón y Cajal, both the telegraph metaphor and Golgi's nerve network did a grave injustice to the nervous system by denying its plasticity, its most essential attribute. A structure whose connections did not change was incapable of learning or growth, two capabilities for which any theory of the nervous system must account. Like Charles Darwin, who represented all of life as a great tree, Ramón y Cajal preferred to represent the organic with other organic structures.

Networks of Nature and Culture

British physiologist George Henry Lewes (1817–78) shared Ramón y Cajal's mistrust of the telegraph metaphor. While he avoided comparing neural "circuits" to technological ones, however, he also rejected the concept of independent nerve cells, supporting the nerve-net hypothesis as the only model that accounted for the nervous system's ability to act as a whole. There is no correlation, then, between belief in the nerve net or belief in autonomous neurons and enthusiastic use of the telegraph metaphor. Ramón y Cajal disapproved of the comparison because he thought it suggested a network, and Lewes disapproved of it because he thought it attributed too many powers to individual cells. Studying the relations among neurons in the 1860s and 1870s, Lewes stands midway between DuBois-Reymond and Ramón y Cajal. In the 1850s, the German physiologist felt that telegraphy accurately illustrated the ways that nerve cells transmitted impulses, but by the 1890s, the Spanish histologist saw the telegraphic "grid" as utterly different from a living communications system. While Lewes embraced the idea of an organic network, he felt that wires, keys, and batteries had nothing to do with the way that the living system worked.

Lewes began his career as a medical student, but by the early 1850s he was primarily a philosopher.[21] Writing Johann Wolfgang von Goethe's (1749–1831) biography revived his interest in comparative anatomy, however, and he admired the great writer's ability to see "Unity in nature" (Lewes, *Life of Goethe* 347). Evaluating Goethe's botanical and anatomical studies, Lewes commented that "while botanists and anatomists were occupied in analysis, striving to distinguish separate parts and give them distinct names, his poetical and philosophic mind urged him to seek the supreme synthesis and reduce all diversities to a higher unity" (*Life of Goethe* 351). As a writer who found himself outside of the scientific establishment, Lewes identified strongly with Goethe.

In 1853, when Thomas Henry Huxley (1825–95) attacked Lewes for his lack of scientific knowledge, Lewes resolved to undertake his own zoological experiments (Haight 195–96; Menke, "Fiction as Vivisection" 620–21). He began by collecting mollusks, about which he published the short volume *Seaside Studies* (1857). His real goals, however, were to understand how the human mind and body interacted and to explain their interdependence in terms accurate enough for experimental scientists but comprehensible to lay readers.

During the 1860s and 1870s, Lewes researched and wrote *Problems of Life and Mind,* a synthetic work that linked neurophysiology, psychology, and ethics. *The Physical Basis of Mind* (1877), the second volume of this study, accurately describes the most recent experiments in neuroanatomy and neurophysiology, which Lewes carefully recorded after years of interviews with the scientists performing them. While Lewes was never a renowned experimentalist like Helmholtz or DuBois-Reymond, he did perform his own physiological experiments (particularly on frogs), and he understood scientists' problems, strategies, and results.

To learn about the latest developments in the natural sciences, Lewes made several trips to Germany, where he met the leading anatomists and physiologists of the day. In 1854, while writing his biography of Goethe, he met Johannes Müller and Emil Dubois-Reymond in Berlin (Haight 171). When he traveled to Munich in 1858, he spent hours discussing chemistry with Justus Liebig (1803–73), and the physiologists there offered Lewes "extensive apparatus and no end of frogs" (Haight 261). In January 1868, he visited Bonn and Heidelberg to discuss science with German physiologists. There is no mention of a meeting with Helmholtz, but Helmholtz was working in Heidelberg at the time. Pleased by the English philosopher's interest in their work, the German physiologists took a great liking to him, and they gladly shared their knowledge.

One of Lewes's most productive trips took place in the spring of 1870. In Berlin, DuBois-Reymond offered Lewes tickets to a university program and introduced him to Carl Westphal (1833–90), one of the leading neurologists in Europe. Westphal took Lewes not just to his laboratories but to his hospital wards to see many "varieties of mad people." Lewes's partner, George Eliot, who accompanied him on almost all of his trips, noted that Westphal was "delighted with George's sympathetic interest in this (to me) hideous branch of practice" (Cross 3: 77). In Vienna, Lewes met anatomist Theodor

Meynert (1833–92), who several years later would teach Sigmund Freud. A painstaking investigator, Meynert spent days showing Lewes his preparations and discussing his findings, while by night, Lewes read the anatomist's studies on the spinal cord (Haight 425). By the time Lewes wrote *The Physical Basis of Mind,* he knew as much as many leading anatomists and physiologists about the ways nerve cells were connected.

Although Josef von Gerlach's article describing a nerve network did not appear until April 1872, Gerlach developed his carmine stain in 1865, and it is very likely that Lewes heard of his observations on one of these trips to Germany. Certainly Lewes knew about Gerlach's work, for he incorporated several of the German neuroanatomist's drawings into *The Physical Basis of Mind.* As a caption to one of them, Lewes wrote, "the processes subdivide into a minute network, in which the fiber also loses itself" (237). Lewes urged his readers to "accept the anatomical fact of a vast network, forming the ground-substance in which cells and fibers are embedded, and with which they are continuous" (340).

While Lewes embraced Gerlach's idea of a nerve network, he remained cautious about adopting any hypothesis without assessing its conformity to anatomical facts. Like Golgi and Ramón y Cajal, he feared the influence of "physiological prejudices." Lewes shared the Spanish scientist's sensitivity to language and his deep awareness that the wording of scientific theories would affect investigators' observations in the laboratory. In *Problems of Life and Mind,* Lewes called scientific models "fictions" and considered them "potent" and "welcome" in science (Menke, "Fiction as Vivisection" 634). Like literature, he warned, they were worthless if they did not reflect external realities, but by definition, as something "made," they could never be identical to the events they represented (Brody, "Physics in *Middlemarch*" 47).

As a writer describing the nervous system, Lewes recognized the affinity of his task with that of his partner, George Eliot, whose novels abound with images of networks and webs. In the 1850s, Lewes's "reinvention" of himself as a scientist paralleled Eliot's recreation of herself as a fiction writer (Menke, "Fiction as Vivisection" 621). To introduce his essay "The Nervous Mechanism" (part two of *The Physical Basis of Mind*), Lewes selected four epigraphs. Three are statements by prominent scientists, but the fourth is a quotation from Eliot's poem "The Spanish Gypsy." While the presence of poetry is interesting in itself, the theme of Lewes's selection is even more significant. The passage reads: "our nimble souls can spin an insubstantial universe suiting our mood, and call it possible sooner than see one grain with eye exact, and give strict record of it" (Lewes, *Physical Basis of Mind* 137). In this scientific essay, Lewes uses Eliot's image of "fictions" spun by the imagination to suggest the "imaginary anatomy" of investigators failing to acknowledge anatomical facts.

Lewes's use of poetry as a source of advice for scientists indicates that he sees science and literature as closely related. While "imaginary anatomy" offends him, he recognizes that anatomy is no pure realm of facts in which an observer's input is a contamination. Nineteenth-century scientists themselves were deeply concerned with terminology and needed no literary figure to alert them to the importance of language in their work. Lewes's writing in *The Physical Basis of Mind,* however, reflects the years spent discussing "connections" with his lifelong companion. While Lewes's descriptions of interwoven tissues helped to inspire Eliot's web imagery, her images of weaving shaped Lewes's "fictions" of the nervous system as well.

Like Helmholtz, Lewes believed that "thoughts differ from sensations as signs from things signified" (Rothfield 92). He was keenly aware of the rift between neural events and the words used to describe them, and he pointed out the distortions that could be created by inappropriate terms. "Nerve force," for instance, was a "convenient symbol" or "shorthand expression" revealing nothing about what it described (*Physical Basis of Mind* 169). Lewes worried that phrases like "transfer of force" would convince readers that an actual substance—like the traditional nervous fluid—was transferred when a nerve impulse was transmitted. Perhaps relying on Tyndall's description of magnetism, Lewes explained that when a magnet "communicate[d] magnetic force to iron," it "altered its molecular condition" and "excited a dormant property" in the iron just as a nerve "excite[d] the dormant property" of contractility in muscle. In both cases, however, "nothing . . . actually passed between them" (*Physical Basis of Mind* 170). Lewes warned his readers that "we must always remember that such phrases [as 'transfer of force'] are metaphors" (*Physical Basis of Mind* 176). Like Eliot, he knew that "we all of us . . . get our thoughts entangled in metaphors," and he reminded readers that scientists' "fictions" affected their visions of the body (Eliot, *Middlemarch* 57).

To Lewes, the most repugnant comparison was the telegraph metaphor, in which the cells figured as batteries and the fibers as cables. A passage from "What Are the Nerves?," an article that appeared in the successful *Cornhill Magazine* in 1862, illustrates the kind of thinking to which he objected:

> In man and all animals alike, masses of grey matter, or cells, are placed at the centre, and nerve fibres connect them with the organs of the body. . . . These groups of cells evidently answer to the stations of the electric telegraph. They

are the points at which the messages are received from one line and passed on along another. But besides this, the cells are the generators of the nervous power. For the living telegraph flashes along its wires not only messages, but the force also which ensures fulfillment ("What are the Nerves?" 160.)

While this metaphor had appealed to physiologists earlier in the century, Lewes found it reactionary and misleading. "The old image still exerts its empire," he complained. "Writers are still found speaking of the brain as a telegraphic bureau, the ganglia as stations, and the nerves as wires. The sensory nerve 'transmits a message to the brain' as the wire transmits a message to the bureau" (*Physical Basis of Mind* 179). The "old image" that so irritated Lewes was the traditional one of a nervous "fluid," a term used not just for nerve impulses but for electrical and magnetic forces. For Lewes, even though the telegraph metaphor invoked a modern invention, it relied on the old idea of the brain as reservoir and the nerves as passive conductors. When one considered the way the nervous system worked, nothing could be further from the truth.

The very concept of the nerve, Lewes believed, was an "abstraction," and he claimed that "only by a diagrammatic artifice can the fiber represent the nerve, and the cell the center" (*Physical Basis of Mind* 183, 300). The points and lines of neuroanatomical circuitry did not exist in the body, only in investigators' imaginations, and Lewes condemned this "imaginary anatomy" (*Physical Basis of Mind* 301). To Lewes, comparing the nervous system to a telegraph network made the nerves seem lifeless and inactive. In his eyes, "nerve force" was not something that the nerves carried but something that they *did,* and depicting them as passive did the entire system an injustice. It is interesting that Lewes and the German physiologists disagreed about the telegraph metaphor when they agreed about the active role of nerve fibers and the value of connections. As in Ramón y Cajal's case, it was probably telegraphy—not nerves—that they envisioned differently. In 1877, Lewes pictured an extensive grid of dead wires, whereas Helmholtz and DuBois-Reymond had seen a living, growing network transmitting information.[22]

While for Lewes the telegraph metaphor always remained objectionable, the most misleading trope of nineteenth-century neuroscience, in his opinion, was not a metaphor but a "synecdoche." This figure represented the nerve cell as a "center writ small" (Lewes, *Physical Basis of Mind* 223, 234). Always an admirer of Goethe's scientific ideas, Lewes felt that the "besetting sin" of contemporary scientific analysis was its failure to perform a reconstructive synthesis once it had broken a system down into its component parts. Inspired by Theodor Schwann's cell theory (1839) and Rudolf Virchow's cellular pathology (1858), scientists had begun taking the part for the whole, attributing the powers of an entire organism to individual nerve cells. In the case of the nervous system, they had started to neglect organs for tissues and tissues for cells (Lewes, *Physical Basis of Mind* 299).

One of the highest goals of *The Physical Basis of Mind* was to overthrow this "superstition of the nerve cell" (*Physical Basis of Mind* viii). Like Golgi, Lewes acknowledged that nerve cells existed, but he did not believe that anyone could attribute the function of an entire system to any one part of it. At its worst, such thinking degenerated into "the monstrous hypothesis of particular nerve cells being endowed with thought, instinct, and volition" (*Physical Basis of Mind* 174). Writing that "the cell has usurped the place of the tissue," Lewes asserted that it was simply wrong to look on the cell as the "supreme element" of the nervous system when so many other structures were involved (*Physical Basis of Mind* 224).

Lewes's tone becomes polemical when he attacks "the superstition of the nerve cell," so much so that one begins to suspect the idea repels him for reasons other than its scientific inaccuracy. Lewes calls cell-oriented theorizing "a sort of fetishistic deification of the cell invest[ing] it with miraculous powers," and he compares such thinking to "statements made by savages respecting the hidden virtues of sticks and stones" (*Physical Basis of Mind* 248). For Lewes, as for Golgi, the nervous system could be understood only as an organic whole. No individual element of it could take credit for its activities, which could be carried out only by the entire system. Consciousness, in particular, was equally distributed throughout (Menke, "Fiction as Vivisection" 635). Although few reputable nineteenth-century scientists would have claimed that individual cells performed complex mental functions, Lewes found the idea so repugnant that he spent pages railing against it.

In July 1869, when George Eliot was beginning to write *Middlemarch*, Eliot and Lewes visited the British museum, where they were "enchanted with some fragments of glass . . . with dyes of sunset in them" (Cross 3: 70). Eventually, this changing play of light upon a complex pattern of fragments would become one of the "germinal images" of Eliot's novel (Andres 854). The image proved as amenable to scientific as to literary writing, however, for Lewes also incorporated it into *The Physical Basis of Mind*. In Lewes's book, the fragments of glass became a "kaleidoscope" metaphor used to explain how the nervous system works:

> We must conceive of these [neural] elements as capable of variable combinations, like the pieces of colored glass in a kaleidoscope, which fall into groups, each group having its definite though temporary form. The elements

constitute really a continuous net-work of variable forms. Each action demands a definite group of neural elements, as each geometric form in the kaleidoscope demands a definite group of pieces of glass; but these same pieces of glass will readily enter into other combinations. (155)

Like Golgi, Lewes understood the nervous system as a network in which each element was connected to every other cell. A complex function like a memory, corresponding to one kaleidoscopic image, occurred when a particular combination of elements formed a pattern. These same elements could create any number of patterns, however, and thus perform any number of functions depending upon the combinations in which they acted. When the narrator of *Middlemarch* declares that memory "shifts its scenery like a diorama," she is articulating Eliot's, Lewes's, and Golgi's understanding of the mind (361).

As a physiologist, Lewes looked for functional arguments that would support Gerlach's and Golgi's anatomical evidence for a net. Like DuBois-Reymond, Lewes relied on Faraday when he described how a nerve transmits signals, but he also drew on contemporaneous physical studies of particles and waves. Communication of force in the nervous system, he wrote in *The Physical Basis of Mind*, meant the communication of vibrations, for "the molecules of the nerve are incessantly vibrating, and with varying sweep" (170). According to Lewes, the "nerve force" behaved like a "wave of molecular movement," with the result that "waves of molecular movement due to each stimulus must sometimes interfere and sometimes blend with others" (*Physical Basis of Mind* 169, 326).

Like other scientists of his day, Lewes rejected the idea of action at a distance. While the nervous system involved no vacuums, it did involve the transmission of physical forces to which gaps presented insuperable obstacles. Twenty years later, Ramón y Cajal would speculate that one neuron could chemically influence a nearby cell, but Lewes agreed with most other scientists of the 1870s that there could be no propagation of signals without "continuity of substance" (*Physical Basis of Mind* 177). This physiological requirement made him welcome Gerlach's evidence for a net. Ironically, the telegraph he rejected as a metaphor may also have led him to picture uninterrupted lines of communication (Dierig 56).

Like Golgi and Ramón y Cajal, Lewes believed that "all function depends on connection" (*Physical Basis of Mind* 153). As the Spanish scientist would later point out, the relative influence of a neural structure depended on its ability to combine with other elements in diverse ways. Lewes anticipated Ramón y Cajal's contention that "the physiological rank of a center is . . . the expression of its power of fluctuating combination" (*Physical Basis of*

Mind 155). The more versatile the structure, the more influential it was, for sensation, thought, and volition depended on a "complex physiological web" in which the nerves were the "connecting threads" (*Physical Basis of Mind* 159, 216). Just as Faraday had rejected the idea of isolated particles separated by a vacuum of space, Lewes argued that if nerve cells were isolated, it would be impossible for them to carry out neural functions. To Lewes, it was incomprehensible how unconnected cells could create thoughts or memories, which were always the products of the entire system. "There is no consciousness," he wrote, "unless the whole organism is involved" (*Physical Basis of Mind* 190).

Like Golgi, Lewes argued for a nerve net from two directions at once, suggesting an attraction to the idea that preceded his experiences in the laboratory. With regard to physiology, Lewes reasoned that since higher mental functions were established facts and only connected cells could give rise to such functions, then there must be a net. With regard to anatomy, he relied on other scientists' observations of networks to argue that "whatever is organically connected cannot functionally be separated" (*Physical Basis of Mind* 310). Physiological facts demanded that nerve cells be physically continuous, and anatomical ones demanded that elements be functionally interrelated.

Like Golgi and Ramón y Cajal, Lewes saw himself as a scientist committed to painstaking, unbiased anatomical observation. But given that each scientist observed carefully while thinking physiologically, what accounts for their different views of the nervous system? Born in 1852, Ramón y Cajal was a full thirty-five years younger than Lewes, but Golgi (born in 1843) was also a generation younger. Cultural developments and technological changes cannot account for their divergent opinions.

When viewed from a broad perspective, their understandings of the nervous system were not so very different. All three scientists believed that nerve cells were interconnected in a highly complex manner, the great number and diversity of their connections allowing cells to work in groups in order to carry out higher mental functions. What these scientists disagreed about was the appropriateness of a "fiction"—about whether the term "network," with all of the cultural images it conjured, was the correct word to describe the complex relations among nerve cells.

Lewes's advocacy of this "fiction" can be traced in part to his organicism and his belief in "sympathy" as a civilizing force. By the 1870s, the nervous system had long been the focus of philosophers seeking a scientific grounding for their notions of sympathy. In the eighteenth century, Scottish physicians had given the nervous system a central role in their medical theories

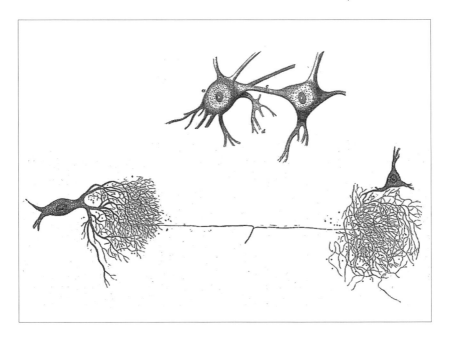

George Henry Lewes's illustrations of conjoined nerve cells.
Above, two anastomosing nerve cells, as shown by Gratiolet.
Below, "Supposed union of two nerve cells and a fibre. The
processes subdivide into a minute network, in which the fibre
also loses itself." The drawing is Joseph von Gerlach's. No
source is given for the Gratiolet or Gerlach illustrations.
(From George Henry Lewes, *The Physical Basis of Mind*
[London: Trübner, 1877], 266, 268.)

because of its integrative function (Lawrence 24). What made the nerves so
interesting to these physicians was that they permitted both sensibility—sensitivity to one's external environment—and sympathy—"communication of
feeling between different bodily organs" (Lawrence 27). Supposedly these
capacities varied from person to person, and aristocrats used references to
their more sensitive nerves to justify their dominant role in society
(Lawrence 20). By the mid-eighteenth century, the quality of sympathy had
come to refer not just to communication within the body but to communication between individuals in society. Those who could feel sympathy for
others were considered highly civilized, and it was believed that such "social
bonding increase[d] with the progress of society" (Lawrence 32).

This respect for sympathy as a civilizing force continued throughout the nineteenth century, even as the nervous system appropriated functions once attributed to bodily sympathy. Both Eliot and Lewes viewed sympathy as the key to moral awareness. Sympathy—the ability to feel what another individual is feeling—is as essential to Lewes's nerve net as it is to Eliot's novels. When Lewes writes that "each sensation and each motion really represents a change in the whole organism," his thinking reflects the eighteenth-century understanding of the nervous system as one that linked the body's organs (*Physical Basis of Mind* 319). His view of the nervous system as "that which binds the different organs into a dynamic unity" makes it the ideal physiological correlate of a society in which sympathy moderates the actions of every individual (*Physical Basis of Mind* 313).

As an organicist, Lewes saw society as a dynamic, complex system in which all elements were interdependent. On a moral level, people could act for the good of this body only when they considered their connections to all of its other members and pictured the effects of their actions on other "cells." Both Lewes, a scientist, and Eliot, a novelist, relied on images of webs because their moral philosophy made the network the best possible "fiction" to describe relations in the body and society. By arguing for a network in the nervous system, Lewes legitimized his model of society as one offered by nature.

Conclusion: The Risks of a Networked World

Like Lewes, network builder Samuel Morse saw the bonds provided by communications systems as a civilizing force. By the early 1850s, Morse's simple, efficient electrical telegraph had been adopted by most European nations, leading to the rapid spread of electronic webs in every direction. Believing that these new channels for communication would transform societies worldwide, Morse predicted in 1835 that "magnetism would do more for the advancement of human sociology than any of the material forces yet known" (2: 46). Morse's greatest hope for the telegraph was that it would prevent future wars by promoting communication among people with different perspectives. In 1855, he prayed that his telegraph would "bind man to his fellow man in such bonds of amity as to put an end to war" (2: 345). Instead, it became the most crucial tool in the wars of the 1860s.

When Morse visited Berlin in 1868, he found himself embraced by grateful officials who showed him how important his telegraph had been in their campaign against the Austrians. While the Germans overtly supported Morse's view of the telegraph as "a means of promoting peace among men," the postal director proudly explained to him how the crown prince and Prince Friedrich Karl had informed each other of their armies' positions, "always over Berlin" (2: 462–63). The demoralized Morse noted that the man kept repeating this phrase. To this official, who worked for the same government as DuBois-Reymond, a telegraph net centered in Berlin was the perfect instrument of imperial power (Carey 212). The new web promoted communication to an extent never before seen, always in the interest of central authority.

Michel Foucault proposes that the "web" or "network" provides an appropriate model for understanding the way power operated in nineteenth-century institutions. In their schools, factories, bureaucracies, and armies, nineteenth-century citizens influenced one another through innumerable cross-links even as they were observed and controlled by others nearer the center (Foucault 170, 228). While all people stood to benefit from a highly interconnected communications net, the greatest benefit would fall to those who assumed the position of the brain, controlling and monitoring the messages moving throughout the system. A sophisticated web could promote the status quo by dampening local fluctuations, particularly when central authority anticipated them with the intention of silencing them in their early stages.

Rather than crushing preexisting identities, however, these networks of power created individuals by defining them in new ways (Foucault 217). It would be a distortion to view nineteenth-century networks as either restrictive, centralized systems for controlling all surrounding activity or liberating devices offering unlimited opportunities for bonding and communication. The networks experienced by nineteenth-century writers both liberated and controlled individuals, shaping their identities through the connections that they provided.

As can be seen from DuBois-Reymond's spiderweb metaphor, scientists' physiological and anatomical studies served these power networks by presenting them as natural formations. The great strength of the Prussian telegraph net, DuBois-Reymond proposed, was that it reflected a well-adapted organic structure. At the same time, cultural images affected the way that anatomists and physiologists viewed the body. Josef von Gerlach saw a web in neural tissue in 1872, a year after the unification of Germany and the centralization of power in Berlin.

For Diderot, in the eighteenth century, the web metaphor had explained how the body maintained control and kept itself informed. While he envisioned a centralized command center, he emphasized the structure's ability to gather intelligence over its capacity to issue commands, and he never clearly associated individual identity with the central point. Golgi's web, like Diderot's, existed to unify a complex system. Proposed more than a century later, however, it differed markedly from the French physiologist's in the role it assigned to cross-connecting links. For Golgi, the body's communications system worked as well as it did because it maximized the number of connections among its elements. A structure in which each cell was joined only to the center could never account for the nervous system's highly complex behavior. Writing during the era of national unification, Golgi believed that only a network could explain how a system functioned as a whole. The fact that Lewes supported the neural-network model in the 1870s further suggests that scientists' images of social bonds affected the way they saw those in the body. The highly visible networks of the telegraph and the railways offered an accessible model for anyone trying to understand how vast numbers of individuals could keep themselves informed and coordinate their activities.

Ramón y Cajal's rejection of reticularism in the late 1880s, however, proves that cultural influences never fully account for what scientists see. The histologist's description of telegraph wires as a "grid" (a word that Gerlach, Golgi, and Lewes never use) indicates either a major change in cultural perception or the action of an original mind. For the same reasons Golgi favored a net, Ramón y Cajal favored a model based on the free association of independent cells. Physiologically, this explained the nervous system's tendency to maximize connections so that cells worked together in groups; anatomically, it was what he observed. For Ramón y Cajal, any viable model of the body's communications system must feature not just complex patterns of connections but the *right kind* of connections, those that could be formed and broken at any time depending upon the body's activities. Could a web represent a cross-connected system that was not just complex but *dynamic?* The Spanish histologist did not believe that it could.

Two decades before Ramón y Cajal overthrew the reticular model, George Eliot created her own "fiction" of a communications net. Like the Spanish scientist, Eliot carefully thought through the structures that might be used to represent communication, and she envisioned complex patterns of connections that could change from moment to moment. For Eliot, whose "particular web" took shape as Gerlach was seeing networks in the brain, the reticular image could explain ongoing changes as long as one imagined a fluid, fluctuating mesh. Her consideration of dynamism in the early 1870s suggests that creative writers, like scientists, base their models both on cultural images and on original visions. Like the anatomists, Eliot thought through society's "fictions" of communication and reworked them to suit her own purposes.

Chapter 3
The Webs of *Middlemarch*

In 1865, in a review of William Lecky's *Rationalism in Europe,* George Eliot (1809–80) referred to "railways, steam-ships, and electric telegraphs, which are demonstrating the inter-dependence of all human interests, and making self-interest a duct for sympathy" ("Influence of Rationalism" 46). To Eliot, railways and telegraphs represented the best of what rationalism could achieve. By facilitating communications, these growing networks increased people's awareness of their connections to others. If people could see themselves as cross-points in a social network instead of as isolated particles, they might overcome their intrinsic selfishness and develop the higher moral quality of sympathy. In Eliot's eyes, one fostered sympathy through knowledge and experience of others, and England's new communications networks offered both.

The webs of Eliot's *Middlemarch* (1871–72) reflect this advocacy of communication on many levels. Her vast knowledge of natural and physical science helped to create its structure and imagery, and the novel's networks recall both Gerlach's nerve net and Faraday's lines of force. In its plot, *Middlemarch* relies on social and economic connections that bind the living and the dead, the reputable and the disreputable. Threads of gossip link character to character and chapter to chapter, so that the novel's structure reflects the bonds that hold society together (Beer; Shuttleworth; Stwertka). While *Middlemarch* often depicts uncontrolled, unreliable, or obstructed communication, it celebrates the flow of information for the potential that it holds. The novel offers the communications web as an epistemological and moral model, suggesting that the securest knowledge and the finest life are those richest in connections.

In exploring Eliot's representation of communications nets, I do not hope to produce a complete, original reading of *Middlemarch*. Instead, I would like to analyze Eliot's use of web imagery in order to compare her understanding of networks to that of contemporaneous scientists. My study of Eliot's imagery reinforces Gillian Beer's and Sally Shuttleworth's fine readings of *Middlemarch*, both of which relate Eliot's depiction of society to her knowledge of science. I will argue that Eliot, like the neuroscientists of the 1870s, saw webs in their tissues not because the scientists influenced her but because they all internalized the humming, ever-expanding communications networks that they saw functioning around them.

The Problem of "Influence": Eliot and Contemporary Science

Eliot's knowledge of science was formidable. Always self-motivated, she systematically studied the most challenging scientific texts of her day and undertook an "intensive self-education in physics" when she was only in her early twenties (Brody, " 'Pier-Glass' Image" 58). Eliot's biographer, Gordon Haight, reports that she "read widely in mathematics, astronomy, chemistry, geology, entomology, phrenology—whatever she could lay hands on" (29). While staying alone in Geneva in 1849, she went twice a week to physics lectures at the Athénée, including one by A. A. de la Rive (1801–73), the inventor of electroplating (Haight 77). Electrochemistry especially interested her because of the relationships it revealed among different kinds of natural forces.

Eliot greatly respected Michael Faraday's imagination and achievements, and his discussions of the interdependence of physical forces intrigued her. In 1850 she attended his lecture at the Royal Society on the magnetization of oxygen (Haight 82). When Eliot published her first novel, *Scenes of Clerical Life* (1857), she requested that copies be sent to eight prominent writers, among them Michael Faraday. The physicist wrote to thank her, and Lewes claimed that she "treasured Faraday's letter beside that of Dickens" (Haight 251). Haight believes Lewes exaggerated and finds her choice of Faraday "most difficult to account for" (251, 246). It is not at all difficult to explain, however, if one considers all the discussions of influences, forces, and lines in Victorian science and culture.

As editor of the *Westminster Review* in the 1850s, Eliot not only solicited articles covering all areas of science, but "her responsibility [was] to link the contributions of various authors together" (Haight 98). An energetic thinker and a demanding editor, Eliot perused scientific writing with the same critical eye she cast upon fiction. She met and corresponded with her scientific contributors, questioning their arguments and their phrasing as she did those of writers from other fields. One might say that this journal was her first web, in which her extraordinary mind spun the essential connections. It was while editing the *Westminster Review* that Eliot met George Henry Lewes in 1852 (Haight 128).

Eliot and the physiological psychologist Lewes became partners in every sense, sharing all aspects of their research and writing. As two of the most highly respected intellectuals in nineteenth-century British society, they personally knew the most prominent English scientists, including Charles Darwin (1809–82), Herbert Spencer, and physicist John Tyndall. While immersed in their creative and scientific projects, Lewes and Eliot read aloud to each other and grappled with the same scientific and philosophical problems. Eliot's fluency in German proved valuable in negotiating scientific texts, since German scientists dominated neuroanatomy and neurophysiology from the 1840s onward. In July 1871, while Eliot was writing *Middlemarch,* she wrote in her journal, "every evening I read a German book of difficult science to my husband" (McCarthy 811). One such volume (which Eliot reports having read in February 1868) was Helmholtz's textbook on acoustics. Eliot and Lewes owned Helmholtz's popular scientific lectures, in German, as well as DuBois-Reymond's *Animal Electricity.*

By the time Eliot wrote *Middlemarch,* she knew a great deal about communication in the nervous system. Lewes was studying the brain as she was writing the novel, and she took an active part in his investigations, learning so that she could discuss key issues with him. As early as 1855 she had written, "I am trying to fix some knowledge about plexuses and ganglia in my soft brain" (Greenberg 33). On 13 and 14 April, 1870, while Lewes was talking to Meynert in Vienna, Lewes noted in his journal: "went over the brain with Polly . . . lesson on the brain to Polly" (Eliot, *George Eliot Letters* 5: 90). This instruction seems to have fed an ongoing interest, for Eliot wrote that in May 1870, during a visit to Oxford, "Dr. Rolleston dissected a brain for me" (*George Eliot Letters* 5: 100). While Eliot was writing *Middlemarch,* she was hearing of nets, continuity, and the folly of attributing actions to individual cells. Although Lewes has sometimes been credited as the source of Eliot's scientific metaphors, however, they are the products of her own creative and analytical mind.[1] Indeed, it is likely that Eliot knew much more about physics than he did (Brody, "Physics in *Middlemarch*" 44).

As an analytical and creative writer, Eliot thought actively about her use of language, proposing in *Middlemarch* that "images are the brood of desire" (222). In *Fragments of Science for Unscientific People,* John Tyndall wrote that "we observe facts and store them up; the constructive imagination broods upon these memories, tries to discern their interdependence and weave them to an organic whole" (Tyndall 1: 353). Tyndall uses "brood" as a verb, an activity the imagination performs, whereas Eliot uses it as a noun, the offspring of desire. Their assertions are closely related, however, in their stress on the creative work of the imagination. Both Eliot and Tyndall, a successful experimental physicist, believed that good thinking relied on interwoven, interconnected images. Many critics have noticed that there is a "science" to Eliot's imagery.[2] Just as scientists rely upon "fictions," small-scale models that convey their visions of complex systems, Eliot used web imagery to help her readers visualize a social system and the innumerable lines of force that made it work.

Because both science and literature arise from our desires to know and communicate perceptions, it is hardly surprising that within any given culture, these desires should produce some of the same images. Like the scientists, Eliot was inspired by the spreading communications networks she saw around her. Because debates about railways and telegraphs offered so many images of lines and webs, scientists from very different fields began using these images in their struggles to articulate theories of magnetism, electricity, and nerve-impulse transmission. It would be a great mistake to conceive of the scientific "influence" on Eliot as a unilateral flow or to ask where she "got" a particular metaphor. With her writing, her editing, and her inspired conversation, Eliot gave to scientific discourse as much as she got. Her web images, like those of the scientists, emerged from exchanges with other minds.

"A Good Thing": The Growth of the British Railway Network

Eliot's praise of the railway echoes the statements of its strongest advocates. While opponents of the railway system deplored its fragmentation of lands, its supporters presented it as a binding force, a much-needed communications net with endless potential to do good. In a biography of George Stephenson, the self-taught engineer who designed England's first locomotives and lines, Samuel Smiles upholds the railways as "a system . . . to revolutionize the internal communications of the world" (206). He presented Stephenson as a visionary hero who had triumphed over ignorance and greed. As a "great means of social inter-communication," wrote Smiles, "[the railways] are felt to enter into almost all relations between man and man" (iv). Just as Samuel Morse presented his telegraph as a promoter of world peace, Smiles promised that by "bringing nations into closer communication . . .[the railways] may tend to abate national antipathies and bind together more closely the great families of mankind" (v). In 1857 Eliot read the Stephenson biography with "real profit and pleasure" (Jumeau 412).

Like Eliot, Smiles used images of physical bonds to describe the dynamic new web that made people aware of their neighbors. Railroads, he wrote, were "drawing the ends of the earth together," and he predicted that "London [would] probably be connected by an iron band of railroads with Calcutta, the capital of our Eastern Empire (318, iv). Stephenson's son Robert, president of the Institute of Civil Engineers, declared that between 1830 and 1855 the British had laid enough railway lines to "put a girdle round about the earth" (Smiles 516).

It is highly significant that Smiles and other advocates of the railways compared their development to the discovery of movable type. In his 1851 history of the English railway, John Francis called the railroad "a power only to be classed with the invention of printing" (1: 65). Like Charles Babbage, who shared their interest in England's new lines of communication, Smiles and Francis upheld printing as the ultimate civilizing tool. Except for printing, claimed Smiles, no other invention had ever done so much to transform society. The parallel between Stephenson's locomotives and Gutenburg's press, however, runs much deeper than a loose association of the two as civilizing forces. As so many midcentury writers observed, both the railway and the printing press facilitated the spread of information, for they "brought towns and villages into communication with each other" (Smiles 318).

By the late eighteenth century, this communication was sorely needed. After James Watt developed an efficient steam engine in 1769, rapid industrial growth created an enormous demand for coal. Since medieval times, resourceful miners had constructed makeshift tramways to move their ore down from the hills (Schivelbusch 4). Eighteenth-century engineers found that preconstructed tracks laid over short distances allowed horses to pull much heavier loads, speeding the transfer of ore to other modes of transportation. At first, however, a network of interconnecting waterways had seemed to offer the best means for moving coal to factories and goods to markets. During the last three decades of the eighteenth century, engineers dug numerous canals connecting England's navigable rivers, along which coal and factory goods could be moved. The canal system, however, proved insufficient as the volume of coal, cotton, and manufactured goods increased. By the 1820s, it often took longer to transport cotton from Liverpool to Manchester than from the United States to Liverpool, and when the canals froze in winter, "communication was entirely stopped" (Smiles 175).

Frustrated merchants and manufacturers began to advocate a horse-drawn tramway system like that used in the mines, but engineer George

Stephenson spoke of a "wonderful machine that was to supersede horses" (Smiles 185). From the time that the steam engine was invented, engineers had proposed using it to produce movement, and a steam carriage was exhibited in Edinburgh as early as 1787 (Francis 1: 48). It took decades, however, before engineers produced locomotives that could pull heavy loads over significant distances.

From the 1820s onward, railway lines sprang up piecemeal, at first connecting major mining centers with ports and industrial cities and later connecting these cities and lines with one another. The Stockton and Darlington Railway, England's first public line, linked the Darlington coalfields with the port of Stockton eight miles away. When it opened in 1825, its cars were moved by horse rather than by locomotive. Stephenson, who surveyed and designed the line, had not thought that passengers would want to use it, but in October 1825 he added a special car, "The Experiment," exclusively for riders (Smiles 200–201). It proved popular.

The thirty-mile Liverpool-Manchester line, also engineered by Stephenson, was designed for locomotive power alone (*New Encyclopedia Britannica* 28: 794). When it opened on 15 September, 1830, one eminent passenger was struck by an oncoming train and later died, but the railway linking one of England's largest ports with its biggest manufacturing city was otherwise an unqualified success (Smiles 304–7). Thanks to the railway, transport time from Liverpool to Manchester dropped from thirty-six hours to one hour and forty-five minutes, and it was soon carrying twelve hundred passengers per day (Lorenz 21; Smiles 307).

For most investors, the railways proved lucrative from the beginning. The Stockton and Darlington paid 15 percent, and despite periodic "panics" caused by fraud and unchecked growth, businessmen clamored to construct new lines and to have their cities "connected." By the time Helmholtz and DuBois-Reymond conducted their experiments in the late 1840s, British and European railways had come "to resemble a web connecting all significant centers of population" (O'Brien 182).

By the 1840s, burgeoning railway networks had changed the way Europeans viewed time and space.[3] Because most trains moved three times as fast as stagecoaches, cities connected to the network in the 1830s suddenly found that the distances separating them had "shrunk" to a third of their former size (Schivelbusch 33–34). Besides altering the overall significance of spaces, railway construction affected the physical appearance of the land. Built for profit and laid with great rapidity, railway lines altered landscapes and cityscapes, usually for the worse. To minimize friction, British engineers tried to keep their lines level whenever possible. On low-lying ground, they laid tracks along raised embankments, and when they encountered hills, they simply cut through the offending obstacles.[4] The new railway lines were thus painfully visible, and many travelers complained that they had "lost contact with the landscape" (Simmons 19–20; Schivelbusch 23). To many people of the 1830s and 1840s, these lines seemed like slashes across the face of the country. They divided estates, plowed through historic sites, and penetrated ancient city walls (Simmons 155). Disturbing city dwellers as much as country dwellers, the "gigantic geometrical brush-strokes" of engineers threatened to make all places look alike (Schivelbusch 179).

Not surprisingly, the first to oppose the railways were the canal owners, who saw immediately that the new mode of transportation threatened their livelihood. Mobilizing quickly, they circulated pamphlets to turn the public against the new communications system. Locomotives, they warned, would pollute the air so that birds would drop out of the sky. The sparks they threw would ignite the houses they passed, and their boilers would burst, killing all of their passengers (Francis 101–2; Smiles 218). Early opponents of the railway directed themselves particularly to the poor, who by the 1830s were all too willing to believe that technological developments would alter their lives for the worse. Incited by canal owners and by their land-lords, pitchfork-wielding peasants attacked surveyors and drove them out of their districts. In one case, a group of miners threatened to throw a surveyor into a pit (Smiles 177). Surveyors and engineers persisted, in some cases performing their tasks by night, but opposition to the railways remained fierce throughout the 1830s.

The best-organized and most vociferous enemies of the railways were the landed gentry, who warned that the locomotive would be "the greatest nuisance, the most complete disturbance of quiet and comfort in all parts of the kingdom" (Francis 1: 120). When a bill approving the London-Birmingham line reached parliament in 1832, it passed the House of Commons easily, but seven-eighths of the landowners in the House of Lords voted against it. If railroad companies had their way, they argued, "game would cease to be, and . . . agricultural communication would be lost; . . . not a field existed but what would be split and divided" (Francis 1: 174). Many of the landowners' objections reflected their aristocratic stress on inherited memories and traditions—as opposed to commercial success—as the basis of identity and social position. One protester declared that "scenes and spots which are replete with associations of great men and great deeds, cannot be pecuniarily paid for. . . . Homes in which boyhood, manhood, and age have been passed, carry recollections which are almost hallowed" (Francis 1: 186–87). In many cases, hypocritical landowners voiced such objections merely to exact a higher price for their lands, for those who held out stood to win three or four times the figure originally offered (Smiles 331–32; Francis 1: 187–88). Most, however, were genuinely offended at the thought of lines cutting across their estates.

Almost all objections to the railways in England focused on their violation of privacy and private property. Despite dire warnings about pollution and noise, the railways drew the sharpest protest not when they first employed locomotives but when they began to carry passengers (Simmons 146). Open to the public, they brought travelers into contact with one another as they had never been before. In more ways than one, the railways were the great equalizer. Extolling the British railway in 1851, John Francis boasted, "it purchases; it sells; it equalizes prices; it destroys monopoly; it places the poorest tradesman on a level with the wealthiest speculator. . . . it is controlled and controllable by all" (2: 280).

Not everyone, though, approved of the new connections among people that the railways promised to establish. Well-to-do riders feared being shut up in a compartment with a madman or a diseased person (Schivelbusch 80–84; Simmons 16). While the railways may have brought people together physically, they never became the "technical guarantor of democracy" that writers like Francis had hoped they would be (Schivelbusch 70). Unlike the U-shaped interiors of eighteenth-century stagecoaches, which encouraged conversation among passengers, the European railway compartment offered physical proximity that was no longer desirable. Embarrassed, afraid to meet each other's eyes, nineteenth-century middle-class passengers wanted only to reach their destinations as quickly as possible, and they usually read to avoid unwanted interactions (Schivelbusch 73–76). While the railways encouraged recreational and business travel, a cultural demand for privacy kept them from creating the instantaneous bonds of sympathy for which their advocates had hoped.

As Britain's railway network grew, the cultural and economic value of privacy became an issue not just in its design but in its management. By the late 1830s, the rapidly expanding system showed a need for centralized governance, and public opinion demanded that "these lines of communication must be placed under undivided control and authority" (Schivelbusch 28). Despite its resemblance to a living network, the British railway system of the 1830s was not a "unified entity" but a "multitude of individual lines, isolated, working without coordination" (Schivelbusch 29). Nothing could have been further from the organic unity of Diderot's web of nerves.

Just who was to control the railways became a hotly contested issue. In France, the government left construction to private contractors, but the system itself was always regarded as a government matter. British railway companies, however, were originally controlled only by their shareholders in commerce and trade (O'Brien 29). By 1832, the railways had begun outcompeting privately contracted postal carriers at delivering mail, and while debating a host of other social reforms, parliament decided that railway travel was a form of communication and vowed to tax it (Francis 1: 212, 283). For decades, the battle raged. Railway companies argued—as the landowners once had—that "the railroad was formed by private enterprise and maintained by private individuals" (Francis 1: 285). The government, on the other hand, wanted to control what it viewed as a national communications system. Both claimed to have only the public good in mind. In the end, Prime Minister Robert Peel settled the matter by insisting that if the landowners had sacrificed the right of private property for the public good, then the railway owners would have to do the same (Francis 1: 286). The organic web that railway advocates envisioned achieved its unity only through a long, costly struggle against the conflicting cultural values of private property and private space.

In 1849, "Railway King" George Hudson became the focus of public hostility toward the railway companies. Fat, loud, and eager for publicity, Hudson proved the perfect subject for caricaturists and became the first middle-class Briton easily recognizable as the result of a media campaign (Simmons 147). In a cartoon entitled "How He Reigned and How He Mizzled," Alfred Crowquill depicted Hudson at the center of an enormous spiderweb in which a number of flies are struggling. Unlike Diderot's eighteenth-century spider, Hudson appears to be immobilized, perhaps caught in his own web (Simmons 147). His limbs are spread-eagled, and the filaments emerge from his back in such a way that if he moves—if he can move at all—he will destroy the network. Caught in his own web, he resembles the trapped flies more than he does a spider.[5] The caption reads, "He does an extraordinary number of lines!!"

The contrast between Diderot's and Croquill's images suggests the vast differences between eighteenth- and nineteenth-century understandings of networks, both technological and organic. Very much in control of her dense web of nerves, Diderot's spider operates an intelligence system that can be seen as an extension of her own body. Crowquill's drawing, on the other hand, depicts the spider as trapped, incompetent. Where Diderot's web empowers, Crowquill's web restricts. It is a network that is out of control. It is a network that controls.

Like nineteenth-century scientists, Victorian writers observing the growth of communications networks saw both their potential to empower

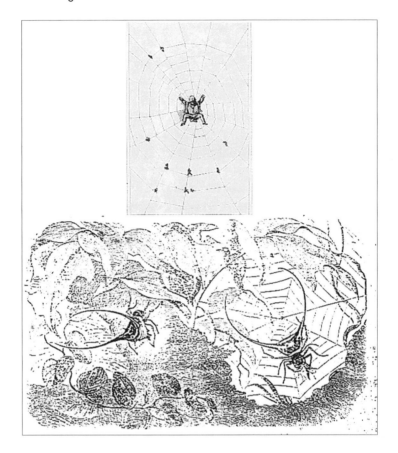

Above, Alfred Crowquill's caricature of George Hudson.
Below, A. E. Brehm's drawing of *Tetragnatha extensa.* (From
Jack Simmons, *The Victorian Railway* [New York: Thames
and Hudson, 1991, Ill. 43] and A. E. Brehm, *Illustriertes
Thierleben: Eine allgemeine Kunde des Thierreichs* [Bil-
burghausen: Verlag des Bibliographischen Instituts, 1869],
6: 580.)

and their potential to entrap. By offering new connections, these webs
promised to liberate individuals who led restricted lives. On the other hand,
the new connections violated private spheres, ensuring that individuals
would be accessible and could be known in disturbing new ways.

While George Eliot depicts the railroad in a more positive light, her writings as a whole reflect Victorian ambivalence toward the railways. As an

eleven-year-old girl, she would have heard local reports about railway workers surveying the London-Birmingham line, and she may have witnessed the surveyors' activity—and local opposition to it—when workers passed through her district in 1831 (Jumeau 411). While working as a civil engineer, Eliot's lifelong friend Herbert Spencer had helped to design this same line (Haight 111). Later in Eliot's life, when her writing had made her wealthy, her business manager, John Cross, invested much of her income in railways and other public utilities (Haight 458).

Eliot's letters and journals reveal the ways in which railway travel was affecting all Victorians' lives. Eliot's journeys with Lewes between London and the countryside were conducted entirely by railway, as were most of their trips through Europe. While she benefited from the travel opportunities that the railways provided, she dreaded the "fatigue and excitement of a long railway journey" (Cross 3: 92). Eliot sympathized with Dickens, who she believed never fully recovered from the railway accident he suffered in 1865. In the spring of 1870, shortly before Dickens's death, she wrote to John Blackwood that "the constant journeys by express train must contribute to the terrible wear and tear of his work" (*George Eliot Letters* 5: 82). Like the physicians of her day, Eliot feared that the jolts and jarrings of the nation's new communications system took a toll on the body's own network of nerves.

She could see that when peaceful resorts were connected to the railways, they were sometimes overrun with people. In 1868, Eliot mourned the way that Torquay was being built up and wrote that "we are afraid of being entangled in excursion trains" (Cross 3: 26). As her popularity grew, she found that isolation could be a blessing, and she rejoiced that her temporary home in Surrey was four and a half miles from any railway station (Cross 3: 74). In August 1871, when she was completing *Middlemarch,* she commented that "the train rushes by every now and then to make one more glad of the usual silence" (Cross 3: 102).

Exhausting and irritating as they could be, however, the railways facilitated the flow of information, a fact all too apparent to a voracious reader like Eliot. Because reading on trains was almost obligatory, bookstalls and lending libraries began offering *romans de gare* from the 1840s onward (Schivelbusch 65). Her own novels were sold in railway stations, and she occasionally commented about books she saw there on display (Cross 3: 51). Like the advocates of the railway, Eliot saw that the growing network promoted the circulation of ideas.

Wolfgang Schivelbusch has proposed that the railways emerged with the capitalist ideology of the industrial revolution, in which the circulation of commodities represented the highest good. Under such a system, he believes, "whatever was part of circulation was regarded as healthy, progressive, constructive; all that was detached from circulation, on the other hand, appeared diseased, medieval, subversive, threatening" (195). Arguments both for and against the British railways support this interpretation. Impoverished workers feared that further technological development would make their lives even more machinelike, a reasonable concern when one considers how rapid industrialization and urbanization had worsened their living conditions. Upper-class opponents complained of the way new lines cut up their lands and violated their privacy. Both aristocratic and proletarian opponents saw railway networks as a fragmenting, shattering, and dehumanizing force. It was the middle class that stood to benefit from the railways economically, that saw them as a creator of life-giving bonds.

Still, views of the railway networks as disempowering or empowering did not fall strictly along class lines but circulated through nineteenth-century European societies. Whether one regarded the new lines of communication as divisive or inclusive depended on one's understanding of subjectivity. For Eliot, who based identity on interpersonal relations, the proliferation of connections could only enhance one's sense of self by making one aware of one's responsibilities to one's neighbors. While fast-paced communication had its drawbacks, Eliot agreed with her character Caleb Garth, known in Middlemarch—in most cases—for his good judgment. The railroad was a "good thing."

The Networks of Middlemarch

Early in the "Berg" notebook for Middlemarch, Eliot wrote, "The highest rate of the nervous discharge is some 32 yards per second. The electric discharge travels at the rate of 280,000 miles per sec. The one velocity is thus nearly 16,000,000 times the other" (Pratt 11).[6] The statement appears under the heading "References for Accent and Quantity" and at first seems out of place among Eliot's notes on ancient medical practices.

Certainly the discovery that nerve impulses traveled so much more slowly than electricity was one of the great shocks of nineteenth-century physiology. In The Physical Basis of Mind, Lewes wrote that the speed of nerve impulses was "not greater than the pace of a greyhound, whereas the velocities of light and electricity are enormously beyond this" (170). Helmholtz's demonstration of this surprisingly low velocity had such repercussions that by the 1860s, both scientists and nonscientists were thinking about the body's system of communication. In 1869, on a trip to Florence, Eliot took time "to see Professor Schiff demonstrate a machine for measuring the speed of thought" (Haight 414). With her extraordinarily active mind, Eliot was interested in everything, but these references to the velocity of nervous action indicate a crucial pattern in her thinking during the years in which she wrote her finest novel. *Middlemarch* is a story of sympathy, networks, and communication, and the movement of information is at the heart of her project.

Middlemarch emerged only gradually, and the process of its composition reveals its intimate connection with physiological issues. In March of 1867, Eliot referred to it simply as an "English novel" and did not "sketch a plan" for it until January 1869 (Pratt xxi). By July of 1869, she was "meditating characters" and writing an introduction, and by October 1869 she had written almost a hundred pages about Lydgate, the Vincys, and the Featherstones (Cross 3: 69–74). Then, on October 19, Lewes's son Thornie died after a long, painful bout with spinal tuberculosis. Lewes and Eliot were devastated, and no progress was made on either *Middlemarch* or *The Physical Basis of Mind* for many months. In March 1870, they traveled to Europe largely to revive themselves, so that their interaction with DuBois-Reymond and Meynert occurred during a critical pause in each of their projects. In December of 1870, Eliot began "experimenting in a story ('Miss Brooke') . . . a subject," she wrote, "which has been recorded among my possible themes ever since I began to write fiction, but will probably take new shapes in the development" (Cross 3: 91). *Middlemarch* as we know it emerged in late December 1870, when Eliot decided to fuse "Miss Brooke" with the Middlemarch story (Haight 432).[7]

For the wrapper of *Middlemarch,* Gordon Haight reports that Eliot's editor, Blackwood, offered her "a design crawling with vines and foliage and meaningless scrolls, which (unhappily) the author liked" (436). Considering Eliot's philosophical leanings and scientific interests, however, it is no wonder that she approved of this tangled bank enfolding her text. The image of interwoven organic growth conveys not just the novel's themes but also its structure and the process of its composition. Eliot's second *Quarry* for *Middlemarch* consists largely of lists on which one notices a recurring heading: "relations to be developed" (Kitchell 45). Like the vines on her wrapper, Eliot's stories and characters gradually grew together.

Eliot's novel, however, can also be represented as a text woven by human hands, and the text's own images and self-references usually present it as a web. Weaving and spinning have served as metaphors since ancient times, and in creating her imagery, Eliot took full advantage of her web's classical roots. Because of the complex interrelations they involve, the female activities of weaving and spinning have long been associated with the creation of fictions, of both elaborate texts and deceitful lies. Well aware of this tradition, Eliot noted in her *Quarry* that in Corinth "there were a thousand female ministrants weaving, embroidering, and seducing" (Pratt 197). Early in the novel, Eliot's protagonist Dorothea finds herself relegated to a world of female stitchery, haunted by "the ghost of a tight-laced lady revisiting the scene of her embroidery" (*Middlemarch* 50). Many critics have noted Dorothea's association with Ariadne, the woman whose connecting thread allowed Theseus to escape the labyrinth (McMaster 112; Shuttleworth 169). With her sympathetic nature, Dorothea struggles to discern the innumerable threads she holds. She tries to understand the relationships that make up her own and her neighbors' lives. Ultimately, she is able to do good not by creating connections but by reminding other characters of those already in place.

The world of *Middlemarch* is one of complex social bonds that link each character to all the others. In this network, the most crucial links are those of family: some unavoidable, some established by choice. The banker Bulstrode, for instance, is married to Vincy's sister. Featherstone was first married to Garth's sister, then to the sister of Vincy's wife. Lydgate becomes part of this web when he is ensnared by Rosamond—whom he attracts because he is "well connected" (*Middlemarch* 62). "Who of any consequence in Middlemarch," asks the narrator, "was not connected or at least acquainted with the Vincys?" (65). In Eliot's web, as in Lewes's and Golgi's nerve networks, the countless connecting links ensure that every event will affect the entire system.

Interwoven with these family ties are economic bonds, and the web of financial obligation reinforces the connections of marriage and blood. Because of his wealth, Bulstrode is linked directly or indirectly to every character in the novel, and he influences everyone in town. His concealment of Will's mother and the subsequent discovery of this secret connect the two main networks of the plot, affecting Lydgate and Rosamond as much as Will and Dorothea. For similar reasons, rich old Featherstone creates vital cross-links. The machinations of his will leave Fred without resources to pay the Garths, and his introduction of Rigg to the Middlemarch network leads to Raffles's uncovering of Bulstrode's past. As in the nervous system, those elements capable of forming the most connections exert the greatest influence, and the wealthier one is, the more threads grow out of one's hands. Mary's remark that "the codicil [of Casaubon's will] had perhaps got mixed up with the habits of spiders" proves all too accurate (414).

Even the power of wealth, however, does not allow characters to escape another system of connections that Eliot depicts, those between the present and the past. For Eliot, the past was always a "quivering part of [one]self," and the webs that she revealed extended through time as well as space (*Middlemarch* 425).[8] Her epigraph to chapter 70—"our deeds still travel with us from afar, and what we have been makes us what we are"—complements that of chapter 4—"our deeds are fetters that we forge ourselves" (485, 22). The former introduces the chapter in which Bulstrode hastens Raffles's death, the latter the chapter in which Dorothea resolves to marry Casaubon. Both decisions are made instantaneously, and both have grave consequences transcending the life spans of the characters involved. In *Middlemarch,* the will of one's forbears combines with one's own past choices to act as a "dead hand" influencing events in the present. Raffles, the blackmailer, asks Bulstrode for an "independence," and his flawed, wealthy victim longs for the same sort of deliverance. Neither can achieve his goal, for each exists only as a nexus in a vast web of humanity. Each can decline to know the other no more than the "droll dog of a thief who declined to know the constable" (*Middlemarch* 366).

The narrator herself takes a detached, often ironic view of these social ties. They are not a cause for celebration; they are simply there, and must therefore be tolerated. Reflecting on Rosamond's enthusiasm for Lydgate's insipid "connections," the narrator remarks: "to most mortals there is a stupidity which is unendurable and a stupidity which is altogether acceptable—else, indeed, what would become of social bonds?" (403). Many social interactions consist of meaningless, time-consuming exchanges, but even this cynical observation offers a faint warning that dull relations are better than none at all. It is pointless, the novel suggests, to deny family connections, no matter how maddening they may be. When Lydgate's and Dorothea's first meeting leaves them both unimpressed, the narrator reflects that "any one watching keenly the stealthy convergence of human lots, sees a slow preparation of effects from one life on another, which tells like a calculated irony on the indifference or the frozen stare with which we look at our unintroduced neighbor" (64). To deny one's links to others is to deny life itself, and those who view social bonds as an obstacle to action must either change their perspectives or fail altogether.

No even web, *Middlemarch* is really a complex of webs imperfectly joined, and its "dominant metaphor" is not these networks themselves but "the *seeing* of the *web* in a certain *light*" (di Pasquale 430, original emphasis).[9] The narrator does not spin the web, which is presented as already existing. Instead, she exposes it as a neuroanatomist might, "analyzing webs of organic relations" (Rothfield 88). It is no coincidence that the narrator's apology—"all the light I command must be concentrated on this particular web"—introduces the chapter on Lydgate's investigations of human tissue (96). To develop the relations that she must, Eliot reveals each character as each is seen by some of the others. The reader learns of key events as the characters do, through imperfect communications and from limited perspectives.

While Eliot's web of social relations is intricately woven, it is also selective, following just some of the consequences that ensue when new connections are formed and old ones are strained. Repeatedly, Eliot warns readers that the portions of the web she reveals are only the merest fragment of what one might see. With their limited views of social bonds, Eliot's characters too often see human relations as constricting and view themselves as Crowquill depicted Hudson, caught in webs that they do not remember having woven. *Middlemarch* abounds with images of yokes, fetters, and harnesses, all of which represent the characters' faulty perceptions of social bonds. Dorothea's sister Celia first appears as a "yoked creature," although it is Dorothea who will soon find herself entrapped (7). If one can believe Casaubon, Will initially sees himself as Pegasus, regarding "every form of prescribed work" as a harness (55).

This imagery of bondage centers around Lydgate, who, like Gulliver, is too long unconscious of his surroundings and awakens to find himself immobilized by Lilliputian threads. At first, he is merely irritated by "the hampering threadlike pressure of small social conditions," but when he begins to see how they affect his own actions, he defends Farebrother by saying, "there are so many strings pulling at once" (124, 343). Lydgate begins as one of those young men who think "that Mammon shall never put a bit in their mouths" but ends by "bowing his neck under the yoke" (97, 411). Ironically, it is because of a chain that he becomes engaged to Rosamond. She drops it, and when he stoops to retrieve it and finds her trembling before him, the narrator tells us, "he did not know where the chain went." Within minutes, he has "bound himself" to its owner (208). Always careful with her words, however, Eliot presents the enthrallment as a deed for which Lydgate is responsible. He is not enchained; he has enchained himself, and if he feels shackled, the fetters are deeds that he has forged himself.

For Eliot, the key to avoiding "yoked loneliness" and "the hideous fettering of domestic hate" was not to avoid connections to others—an unavoidable condition in any human society—but to accept these connections as a vital aspect of one's being (*Middlemarch* 462, 461). In her eyes, connections became fetters only when viewed from a selfish perspective, judged for the personal desires they frustrated rather than for the social relationships they enabled. In her notebooks for *Middlemarch,* Eliot loved to record the ironies of science, one of which was the multiple possibilities of iron bonds. The jailer sees iron as a restriction, she noted, but the electrician "sees in it only a channel of communication" (Pratt 61). Despite Lydgate's perceptions of social ties, the bonds among characters do not impose rigidity. Instead, they permit movement.

In arguing for a physically continuous nerve network, Lewes had proposed an analogy to the circulatory system. "Is it fanciful," he asked, "to regard this network of fibrils as having somewhat the relation of capillaries to blood vessels?" (*Physical Basis of Mind* 308). Like Lewes's nerve net, Eliot's web is a dynamic one (Miller, "Optic and Semiotic" 132). The social connections she depicts are a circulatory system of sorts, permitting the flow of blood, money, and information (Beer 48). It is significant, for instance, that Lydgate becomes interested in medicine when he first observes the valves of the heart. The narrator's account of his response—"he had no more thought of representing to himself how his blood circulated than how paper served instead of gold"—suggests that finance, like blood flow, is a kind of circulation (*Middlemarch* 98; Beer 51). In Middlemarch, the movements of money and blood are interdependent, not just in Lydgate's career but in the very structure of society.

Like a body, *Middlemarch* involves "multiple interrelated systems and circulations," and the communication among them makes it work both as a story and as a living whole (Beer 60). Briefly, the novel invokes two national circulatory systems in the process of development: the railways and the postal service (Beer 52, 57). Like the more traditional networks of family, these systems permit communication, facilitating and maintaining social bonds. It is the much older network of rumor, however, that holds Eliot's novel together, as the reader travels like money or news from group to group. To move the reader from one scene to the next, the narrator often resorts to gossip, *Middlemarch*'s version of the law of cause and effect. In one chapter, the reader hears a stone fall, then studies its ripples in the next. Featherstone's death, for instance, is a key event that affects almost all of the characters. To emphasize the townspeople's interdependence—and their unconsciousness of it—Eliot shows the reader Featherstone's funeral only from a distance, as a backdrop to a conversation among Dorothea, Celia, and Mrs. Cadwallader. The narrator comments: "scenes which make vital changes in our neighbors' lot are but the background of our own, yet, like a particular aspect of the fields and trees, they become associated for us with the epochs of our own history, and make a part of that unity which lies in the selection of our keenest consciousness" (223). With their limited views of social change, individuals often fail to perceive the import of events that will alter not just their own lives but the shape of their society as a whole.

For her novel of provincial life, Eliot purposely selected an era of profound social change, the reform years of 1829 to 1832. Like Ovid, she depicted metamorphoses in progress, exploring their causes and effects, and she carefully studied the political and scientific debates of the period in order to represent these changes accurately. Written between 1869 and 1872, *Middlemarch* looks back with a certain nostalgia to a time in which provincial isolation still existed. By setting her novel in the reform era, Eliot directed her readers' gaze to a time and place when this isolation was just beginning to dissolve. It was during these years that Stephenson and other engineers were planning England's essential railway lines, and the growing network promised to link the most faraway districts to the British nerve center, London (Hooton 194). In this time of upheaval, social and political changes occurred at a rapid pace, yet people were nowhere near as conscious of national events as their counterparts in the 1870s. By 1870, thriving telegraph and railway networks ensured that a character like Bulstrode could never have hidden his past (Beer 52).

As a chronicler of ongoing changes, Eliot was fascinated by the relationships between the local and the global, between the apparently insignificant event and the historically recorded one. In 1829, political and technological changes appeared to be rending the social web, and when viewed from the perspective of local districts, these changes constituted a process of fragmentation. The railroad threatened to "cut up" the country, workers were breaking machines, and parliament was "dissolved" (Eliot, *Middlemarch* 383, 243). From a national perspective, however, the same railway that divided individual estates would unite the country with a powerful communications network, and the riots and debates of 1829 would lead to reforms benefiting the entire national organism. As a study of provincial life, *Middlemarch* is not so much about these reforms as it is about perspective itself. It explores the profound consequences of the local and the trivial and the apparent insignificance of the global and profound.

The Local and the Global

In the 1860s and 1870s, the relationships between individual elements and complex systems became a crucial issue for physicists and biologists as well as for fiction writers. Like the physicist Tyndall and the anatomist Golgi,

Eliot wanted to study how interactions among individuals could produce systemwide effects and how systemic changes were experienced by individuals at the local level.[10] In the 1860s, James Clerk Maxwell (1831–79) was studying how the energetic states of individual molecules might account for the overall behavior of gases. His papers of 1860 and 1866 on the dynamical properties of gases transformed physics by showing that collections of particles could be analyzed using statistical laws (Gillispie, *Edge of Objectivity* 482–85). Ultimately, Maxwell's experiments in statistical mechanics revealed that individual particles and populations of particles followed their own distinct rules. If one knew the conditions to which individuals were subject, one could predict—but never know for certain—how a large group of them would behave (Brody, "Physics in *Middlemarch*" 47–48).

John Tyndall, a frequent guest of Eliot and Lewes from 1865 onward, wanted to learn what held the particles of matter together and how they interacted with the atmospheres of ether around them. In the laboratory, he pursued these questions by changing the temperature (which either increases or decreases molecular motion) and observing changes in the ways that gases absorbed or transmitted light. Tyndall discussed his work extensively with Eliot and Lewes, and Eliot agreed with Tyndall that atoms were real entities, not mere models approximating the structure of matter (Brody, "Physics in *Middlemarch*" 46). Not surprisingly, the physicist's descriptions of nature closely resemble Lewes's depictions of the nervous system and Eliot's representations of social relationships.[11] When one looks carefully, writes Tyndall:

> the idea that nature is an aggregate of independent parts . . . disappears, as the connection and mutual dependence of physical powers become more and more manifest: until [one] is finally led to regard Nature as an organic whole—as a body each of whose members sympathizes with the rest, changing . . . without break of continuity in the relation of cause and effect. (1: 343–44)

Like Eliot and Lewes, Tyndall believed that a complex system could only be understood in its entirety, by considering all of the interactions of its many elements.

When Tyndall wrote about the behavior of atoms, he almost certainly had social forces in mind. Although his popular *Fragments of Science for Unscientific People* may not represent his typical scientific writing style, its essays show his deep awareness that the relationship between an individual particle and a population of particles is a cultural as well as a scientific issue. Often Tyndall's descriptions of particles are highly anthropomorphic. He writes, for instance, that when a gas condenses into a liquid, "the molecules coalesce, and grapple with each other" and that there is a "mutual entanglement of the molecules by the force of cohesion" (1: 386). As was the case with Eliot and nineteenth-century neuroscientists, Tyndall's daily experiences suggested the models he used to make sense of nature.

In writing about the contrary pulls of personal desires and social obligations, Eliot described the interactions among individuals in a strikingly similar way. She explored the same issues as scientists studying molecular kinetics, at times introducing their discourse to her own fictions. Lydgate, for instance, suggests Maxwell's findings when he declares his intention "to be a unit who would make a certain amount of difference towards that spreading change which would one day tell appreciably on the averages" (100). Assessing Will's early feelings for Dorothea, the narrator of *Middlemarch* presents Will's sentiments as a failed scientific model involving both particles and waves: "there are characters who are continually creating collisions and nodes for themselves in dramas which nobody is prepared to act with them" (133). Like a bad scientific model, Will's "fiction" fails because it is based on the perspective of a single element.

Like Lewes and Tyndall, Eliot believed that any good model must take the entire system into account. Selma Brody believes that when Eliot began reading Tyndall, "a whole new area of physics began to appear in her metaphors," but it would be a grave mistake to identify Tyndall as the source of Eliot's images (Brody, "Physics in *Middlemarch*" 46). Eliot and Tyndall undoubtedly influenced each other as writers, but they referred to connections and sympathy because Victorian culture offered them these terms.

In biology, the formulation of cell theory in 1839 had made the relationship between the individual and the collective a central issue. After reading T. H. Huxley's 1853 article on cell theory, Eliot noted in her first *Quarry* for *Middlemarch*: "two fundamental notions of structural and physiological biology: the 1st, that living things may be resolved anatomically, into a comparatively small number of structural elements; the 2d, that these elementary parts possess vital properties . . . and are independent of all direct influence from other parts" (Kitchell 31). Knowing, like physicists, that the complex systems they studied consisted of individual elements, biologists wondered how much they could infer about individual cells based on their observations of large populations of them. Simultaneously, they asked themselves what they could predict about populations, given what they knew about cells in a particular region. At the heart of nineteenth-century debates between neuroanatomists favoring a net and those stressing the actions of individual cells was the problem of independence, of the relationship between the individual element and the whole organism. When Lewes deplored the "superstition of the nerve cell," he admonished anatomists who were failing to see the system's unity; when Ramón y Cajal attacked the reticularists, he criticized scientists who were underestimating the possibilities of individual actions. In *Middlemarch,* Eliot addressed exactly these issues, exploring the ways individual actions created systemwide changes and the ways systemwide changes affected individual lives.[12]

Lydgate's career illustrates how the relationship between local and global events works simultaneously as a biological and a social problem. Applying his belief in independent tissues to his life in Middlemarch society, the young doctor quickly becomes entangled in his own metaphor, mistakenly believing that he can act as an autonomous element in a small provincial town.[13] Hoping to accomplish "small good work for Middlemarch, and great work for the world," Lydgate presumes that by living in the country, he can escape London's political squabbles and reserve more time for his research (*Middlemarch* 102). Middlemarch's new doctor views people and tissues as largely independent, scoffing at Widgeon's purifying pills that "arrested every disease at the fountain by setting to work at once upon the blood" (309). While Eliot suggests that Lydgate's scientific hypotheses have merit, his confidence in his own autonomy becomes a life-destroying premise. In Lydgate's understanding of pathology, all disease can be traced to inflammations of specific tissues, and in Middlemarch, all politics are local. It is no coincidence that Eliot follows her introduction to Lydgate with a chapter on petty politics.

As Eliot presents it, *Middlemarch* is a novel of the small and the trivial, of people and things whose importance eludes one's perception. When the auctioneer Trumbull declares that "trifles make the sum of human things—nothing more important than trifles," he suggests the novelist's own attitude (417). In *Middlemarch,* the intricate connections among people ensure that there can be no inconsequential deed. Even an action dictated by the slightest whim can alter many lives. Rosamond goes riding with Sir Godwin and loses a baby; Featherstone defers changing his will until the last possible moment and leaves Fred—and the Garths—without resources. "Men outlive their love," warns Farebrother, "but they don't outlive the consequences of their recklessness" (356). Sharp-witted Mary best expresses the deadly effects that can be wrought by one false step when she tells Fred of "the ants whose beautiful house was knocked down by a giant named Tom, . . . [who] thought they didn't mind because he couldn't hear them cry or see them use their pocket-handkerchiefs" (444).[14] Triviality and consequence, Eliot suggests, are relative terms. One can speak of importance only from a particular perspective, and her novel is not so much about connections and consequences as about the need to perceive them.

Like Lydgate, many of Eliot's characters fail to see the social and political forces that will influence their lives. When the narrator refers to political events, her descriptions are usually indirect, arising in conversations among characters who are more interested in personal issues. In chapter 3, for instance, one hears about rick burning, machine breaking, and the "sties" in which Brooke's tenants live, but all are subordinated to the more pressing question of whom Dorothea will marry. Despite the squalor of their peasants' dwellings, the Middlemarch gentry see no connection between themselves and these distant rebellions, maintaining their focus on more obvious social bonds.

Eliot makes it clear, however, that local and global politics are inseparable. In Middlemarch, as in Lewes's and Golgi's models of the brain, there is no real independence. Much of the novel's profound irony involves forces that arrogant characters do not see. When Brooke runs for parliament, both the candidate and his potential constituents suffer from limited perspectives. The local weavers feel no more "attached to [Brooke] than if he had been sent in a box from London" (348). While the workers are astute in their judgment of Brooke, Eliot suggests that they are less so in their feeling of alienation from their national capital. Unlike the weavers, Brooke is eager to participate in the national government but cannot see the relevance of his local role. While he "goes into everything," he is a sloppy and stingy landlord, and he fails to perceive how his incompetence at home will undermine his dream of deciding national issues. Will Ladislaw despairs of convincing this "independent" candidate that he must be either for or against the Reform Bill. Even Will, so much more perceptive than his employer, deludes himself when he rationalizes that "the little waves make the large ones and are of the same pattern" (319).[15] Although the local always affects the global, one cannot infer that the local is the global "writ small" any more than one can attribute a memory to an individual cell. Provincial and national changes continuously shape one another, but at the local level, systemwide changes are often eclipsed.

No aspect of *Middlemarch* better illustrates the complex relationship between national and local events than the townspeople's response to the railway. Like the specters of reform and cholera—with which it is juxtaposed—the railway emerges in the characters' gossip as a threat to local order. To the inhabitants of Middlemarch, the network that will link them to the nation appears as a monster ready to divide and consume the town.

"What's to hinder 'em cutting right and left if they begin?" asks Mrs. Waule, and Mr. Solomon fears that the railway is an excuse "for the big traffic to swallow up the little" (382, 383). Landlords and small businessmen worry that the railways will endanger their livelihood, and their workers suspect that the new network, like the canal system, will only be "good for the big folks to make money out on" (386). The residents' fears that "the cows will all cast their calves" and that "railways . . . blow you to pieces right and left" reflect Eliot's reading of George Stephenson's biography as well as her own childhood experiences (*Middlemarch* 382; Jumeau 411). Rich or poor, the people of Middlemarch voice a common fear of losing their identities and being engulfed. While not all of them entertain Hiram Ford's "dim notion of London as a center of hostility to the country," they fear that being connected to this nerve center will destroy the isolation that has helped to create their sense of self (383).

In *Middlemarch,* context is always crucial, and Eliot gives the railway episode a suitably ironic frame. In the preceding chapter, Dorothea envisions a feudalistic utopia, wishing that she could "take a great deal of land, and drain it, and make a little colony, where everybody should work, and all the work should be done well. I should know every one of the people and be their friend" (380). The coming railway, of course, will eliminate any possibility of such an isolated, localized colony. The medieval fiefdom Dorothea pictures, in which she is the beloved central figure, will be replaced by an immense network in which she can derive her sense of worth only from her relations to adjacent subjects.

Because of these connecting links, global changes in complex systems can produce some rather surprising effects at the local level. When railway surveyors pass through the Middlemarch district and are attacked by a hayfork-wielding gang, they precipitate a small incident that for the townspeople proves more significant—in the short run—than their connection to the London-Birmingham line. Seeing the besieged surveyors, idle young Fred Vincy rushes to their defense. By so doing, he catches Caleb Garth's attention, leading Garth to employ him, so that eventually Fred can marry Mary and begin repaying his debt to the Garths. In the eyes of the townspeople, familial and economic bonds constitute the foreground and dominate the visual field. While most characters do perceive some connecting thread between themselves and London, it is a thread impossible to follow, one that quickly vanishes into the distant haze. They fail to see how their personal bonds of marriage, family, and business link them to a much larger network of social ties connecting all members of their society.

A Network of Knowledge

Much of *Middlemarch*'s irony and humor grow out of the characters' "shortsightedness." Dorothea, who dislikes puppies because she fears treading on them, sees far better—and worse—than any other character in the novel. By making Dorothea the focus of the vision issue, Eliot reminds the reader of the many aspects of seeing. Pragmatic Celia points out that Dorothea perceives what no one else can see yet fails to see what to everyone else is perfectly plain. With her sympathetic nature, Dorothea is keenly aware of her neighbors' suffering, but her idealism blinds her to their simple, selfish desires.

Because of her shortsightedness, Dorothea can sometimes intervene at crucial moments and offer other characters valuable help, as when she convinces Rosamond of Lydgate's innocence. The narrator explains that "her blindness to whatever did not lie in her own pure purpose carried her safely by the side of precipices where vision would have been perilous with fear" (257). Ultimately, a blind person wandering along a precipice will fall off of the cliff, advantages of fearlessness notwithstanding. Good vision requires judgment, sympathy, and the ability to compare an object to an ideal template, but it also demands careful scrutiny of the small and the ordinary.

Dorothea always sees the grand social connections binding her neighbors, those of feeling and moral obligation. The threads of petty motives often exert a greater pull, however, and one cannot understand people's interactions without considering these ties as well. Just as a neuroanatomist struggles to resolve tangles of cell processes, Eliot works through the intricate connections in her fiction of social relationships, one that she has developed based on her astute observations of society. Like Golgi and Ramón y Cajal, the realistic novelist is a microscopist, "enamored of that arduous invention which is the very eye of research, provisionally framing its object and correcting it to more and more exactness of relation" (Eliot, *Middlemarch* 113). For both the scientist and the novelist, to see best is to see "exactness of relation," to view structures in context and detail. One can assess the system one is viewing only by studying the connections among its many parts.

When nineteenth-century scientists thought about the structure of knowledge, they conceived of it in terms of the systems they were observing. The way they acquired knowledge shaped their understanding of what knowledge was. Like Eliot, Hermann von Helmholtz made extensive use of web metaphors, but in his writing they were primarily epistemological. In his popular essays, Helmholtz used images of interconnecting threads not just to represent a communications network but to describe the knowledge it made possible. For Helmholtz, knowledge was something that was spun, something constructed by establishing connections. In school, he recalled, "I had a bad memory for disconnected things" (*Science and Culture* 383). He could remember and make sense of facts only after he had determined the general laws that governed them, and he declared that "intellectual satisfaction, we obtain only from a connection of the whole" (*Science and Culture* 97). Lacking the instinct of spiders, scientists often pursued lines of reason and conducted series of experiments without knowing where their threads would lead. A filament of thought originating in a dead-end problem might provide the insight required to solve one more pressing, and a filament associated with one field of knowledge might lead unexpectedly to another. In his lecture "On the Interaction of Natural Forces" (1854), Helmholtz declared that "the thread which was spun in darkness by those who sought a perpetual motion has conducted us to a universal law of nature" (*Science and Culture* 43). Describing his own discovery of energy conservation, Helmholtz wrote that "here the thread is begun to be spun which subsequently led a physician to the law of the conservation of force" (*Science and Culture* 315). The great beauty of science was its ability to reveal unexpected connections in the intricate web of nature. If scientific knowledge was a network, it reflected the systems it explored.

Helmholtz found the weaving metaphor appropriate not just for describing the construction of knowledge but for representing the connections among the forms of energy he sought to understand. He compared natural forces to the threads of a web in order to show that the slightest pull on any one string must affect the entire system. From the time that he first began his studies of animal heat, Helmholtz had suspected that heat, electricity, and movement were interrelated, and he struggled to design experiments that would reveal the threads leading from one form of energy to the next. "The problem," he explained, "was to find, in the complicated net of reciprocal actions, a track through chemical, electrical, magnetical, and thermic processes, back to mechanical actions" (*Science and Culture* 26). Electrical currents intrigued him, for, he discovered, "the application of electrical currents opens out a large number of relations between the various natural forces" (*Science and Culture* 122).[16]

As had been the case with Helmholtz, Eliot's respect for organic relations led her to represent knowledge itself as an interconnected whole. In *Middlemarch*, one learns by discovering bonds of which one was previously unaware, and the structure of knowledge reflects the process of its acquisition. To know, for Eliot, is to understand the relations of cause and effect, but like nineteenth-century scientists, she unfolds the traditional chain of causality into a two-dimensional network of interactive forces. In her novel, she attempts to show "the gradual action of ordinary causes rather than exceptional," creating a field in which no event can be explained by a single cause (Cross 3: 99; Miller, "Optic and Semiotic 133).[17]

When Eliot writes about perception, causality, and knowledge, her images resemble physicists' representations of natural forces. To convey the complexity of social interactions, for instance, she asks readers to picture the movements of electrolysis. In the introduction to chapter 40, the narrator explains that "in watching effects, if only of an electric battery, it is often necessary to change our place and examine a particular mixture or group at some distance from the point where the movement we are interested in was set up" (275). The chapter is preceded by the peasant Dagley's defiance of his landlord, Brooke, and it is followed by Raffles's chance discovery of Bulstrode's letter. In chapter 40, Mary will learn that she need not move away to teach school because Chettam wants her father to manage his estate. In its context as well as its content, Eliot's chapter is a study of causes and effects.

When the narrator refers to the electrochemical system, she admits that her analogy is reductive, asking us to consider the effects "if only" of an electric battery.[18] She makes the comparison, she explains, because of one key similarity between ions in an electrolytic solution and characters in provincial society. In each case, to understand what one is seeing, one must observe not merely what is happening around the electrodes or central characters but what is happening at distant points throughout the solution. Ideally, one would like to make measurements at all points, observing from all perspectives simultaneously. Because this is impossible, both in physics and in fiction, one must settle for taking readings at a few sample points. From these observations, the reader or scientist can derive a general picture of the movements of ions.

Like the images of Helmholtz and Lewes, Eliot's metaphor implies that true knowledge is always knowledge of an entire system. One can understand the relationships between causes and effects only when one sees them played out over an entire web. To introduce chapter 64, which describes Lydgate's and Rosamond's struggle for dominance, Eliot writes in her epigraph, "All force is twain in one; cause is not cause unless effect be there" (447). In *Middlemarch,* causes are meaningful only relative to their effects, and the ties linking each event to its causes and consequences create the lines of force along which Eliot's characters move.

Most of the characters, however, are unaware of these lines of force. Their perception and knowledge consist of incomplete segments, and one senses Eliot's respect for organic, systematic thinking when reading her descriptions of characters who think badly. The narrator's statement that "our passions do not live apart in locked chambers" and Will's declaration that Rome "save[s] you from seeing the world's ages as a set of box-like partitions without vital connection" express Eliot's understanding of consciousness as open, dynamic, and interconnected (114, 147).[19] For Eliot, thought means communication, the flow of energy through the chambers of the mind, and those characters whose stream of thought is obstructed will hurt themselves and others. Bulstrode, for instance, is consistently guilty of "broken metaphor and bad logic of motive" (471). When he rationalizes his greed, "the fact [is] broken into little sequences, each justified as it came by reasonings which seemed to prove it righteous" (427). In depicting the mind at work, Eliot expresses the ideology of circulation (Schivelbusch 195). True knowledge requires open communication and continuous movement, and where there is no circulation, there will be stagnation and disease.

The characters most guilty of fragmented thinking are Rosamond and Lydgate, whose inability to communicate creates a life of marital hell. Their "total missing of each other's mental track" suggests a failure to form connections as effectively as England's new technological communications systems (405). Unfortunately, the narrator reflects, Rosamond's incomplete vision of causes and effects is all too common. She sees no web of causality, only a one-dimensional link between what she desires and its most likely source. "To see how an effect may be produced is often to see possible missings and checks," comments the narrator ironically, "but to see nothing except the desirable cause, and close upon it the desirable effect, rids us of doubt and makes our minds strongly intuitive" (531). Lydgate, the scientist, reasons no better than Rosamond when it comes to personal relations, thinking of "himself as the sufferer, and of others as the agents who had injured his lot" (509). Rosamond is perceptive when she suggests a link between his scientific activities (examining "bits of things in phials") and his social blunders (316).[20] While Lydgate views social conditions as "hatefully disconnected with the objects he cared to occupy himself with," Rosamond has learned to see only those social bonds that she can use to her advantage (407). Both are unaware of the connections binding them to their neighbors, most particularly to each other. Thinking and perceiving in incomplete sequences, they feel "as if they were both adrift on one piece of wreck and look[ing] away from each other" (522). The advent of Dorothea's sympathy saves their marriage but not their happiness.

Marital misery, Eliot suggests—like most other varieties—has its roots in incomplete knowledge. In the summer of 1869, while beginning *Middlemarch,* Eliot wrote in her *Quarry* that for the ancient Greeks, "matters divine, heroic, and human were all woven together . . . into one indivisible web, in which the threads of truth and reality . . . were neither intended to be, nor were actually distinguishable" (Pratt xxxv). For Eliot, such an indivisible, cross-connected structure represented the finest "fiction" and the highest knowledge, a web woven with observations and creativity. In comparison, the "pilulous smallness [of] the cobweb of premarital acquaintanceship" betrays its dearth of vital connections (*Middlemarch* 13). No web amounts to much when it is collapsed, but the particularly low volume in the case of courtship suggests that substantial knowledge requires a great number of connecting lines. To know is to understand the effects that one's actions will produce as they shake each thread of an intricate web.

A Network of Language

It takes considerable effort, though, to visualize these lines of cause and effect. Usually, one must draw them oneself, connecting any fragments of information that one can perceive. If knowledge is a web that one weaves, how is it constructed? How does one form the vital connections, and how does one decide where to place the cross-connecting links? Philologist Max Müller, whom Eliot greatly admired, compared the science of comparative anatomy to the study of ancient languages. In both cases, Müller believed, the scholar had to "learn to make the best of his fragmentary information" (Shuttleworth 14). In *Middlemarch,* Eliot's descriptions of learning reveal the essential role of language in all knowledge. To understand is to connect words and phrases, and one's knowledge is a web spun from the language one has heard.

To a large degree, the network in which Eliot's characters function is a community of language (Shuttleworth 168–69; Miller, "Roar on the Other Side" 243). The district of Middlemarch is a "huge whispering-gallery" in which information reverberates with a perverse tenacity (*Middlemarch* 284). Just as trains and nerve impulses move along their respective networks, allowing complex, highly diversified organisms to function, language moves through the social network, reinforcing existing connections and stimulating the formation of new ones. The communication it makes possible maintains the sympathetic links that hold society together. For Eliot, words were both the nerves and the information they carried, and in *Middlemarch,* language is a network in itself. Messages strengthen or strain social bonds, and language is the stuff of which these bonds are made.

As Eliot's novel makes clear, using language to create meaningful structures is no easy matter. Gossipy Mrs. Taft is "always counting stitches and gather[ing] her information in misleading fragments caught between the rows of her knitting" (182). Passing on what she acquires without thinking it through, she incorporates bits of news as haphazardly as she picks them up. Prudent Caleb Garth, in contrast, waits too long to speak, struggling to connect the scraps of language in his consciousness. "In his difficulty of finding speech for his thought," describes the narrator, "he caught, as it were, snatches of diction which he associated with various points of view or states of mind" (284). While Garth tries desperately to turn these snatches into coherent thoughts, irresponsible Brooke never even makes the effort, failing to realize that knowing means connecting the scraps of information one has heard. When he tells his niece Dorothea that "young ladies are too flighty," she suspects that "the remark lay on his mind as lightly as the broken wing of an insect among all the other fragments there" (11). For Eliot, active thought means the continuous incorporation of new language into a living structure. Words are nothing but broken insect wings if one cannot work them into a dynamic web.

While Dorothea views Brooke's mind as a collection of fragments, it is actually a mind that makes too *many* connections. Brooke's problem is not the inability to forge links but the inability to forge links based on causal relationships rather than simple contiguity. His rambling campaign speech and his imprudent pen illustrate language's ability to spread metonymically, "evolving sentences . . . before the rest of his mind could well overtake them" (*Middlemarch* 201). While knowledge and society depend on strong associations, Eliot implies, not all of these connections are valuable in themselves. Connections create meaning only when they result from open-minded observation, when they are carefully considered, and when they provide vital links that nourish an organic body of knowledge.

Casaubon's pitiful research project illustrates the inviability of a knowledge network built from poorly connected scraps. From the outset, Eliot presents his work as fragmented and lifeless. It is a web, of sorts, but it is one whose threads reflect personal whims rather than true sympathetic links (Cosslett 85–87).[21] Casaubon's untested associations, "as free from interruption as a plan for threading the stars together," offer nothing but glimpses of his own limited perspective (*Middlemarch* 332).

In its goals, Casaubon's quest for the Key to all Mythologies reflects other nineteenth-century searches for origins. Like many romantic philologists, Casaubon seeks the "Ur" form of myths, the original, quintessential version from which all subsequent ones have been derived. Goethe, whose scientific work Eliot and Lewes knew intimately, had envisioned an *Urpflanze,* a primordial plant from which all modern plants had developed. For comparative anatomists and linguists, knowledge of such original forms promised to make all others "luminous with the reflected light of correspondences" (*Middlemarch* 14). Whereas Goethe's botany celebrates diversity, Casaubon regards growth and development as "corruptions of a tradition originally revealed" (14). Eliot's aging scholar hopes "mentally to construct [mythology] as it used to be, in spite of ruin and confusing changes" (9). Casaubon's antiorganic approach produces no coherent structure, for he opposes the evolution and vital processes that make knowledge possible.

Dorothea's feeling of "stupendous fragmentariness" in Rome reveals her growing understanding of her husband's mind (134). Her vibrant presence invites one to read his intellectual impotence as a sexual deficiency, just as his neighbors associate his unhealthy mental attitudes with his obvious bodily defects. Dorothea's suitor Chettam reads Casaubon's lack of vitality as a lack of "good red blood," and Mrs. Cadwallader adds, "somebody put a drop under a magnifying-glass, and it was all semi-colons and parentheses" (47). Like Casaubon's thoughts, she hints, his blood consists of punctuation alone, markings that divide and interrupt the vital flow of language.

Instead of producing new connections or new growth, Casaubon's work consists entirely of notes and "parerga," scratchings that interrupt and qualify other scholars' work. Casaubon never finds his "key," and with his lack of sympathetic connection to the people around him, it is unlikely that he would know what to do with it if he did. For Eliot, the key to scholarship is communication in every possible sense: interaction, movement, contact, and flow. Appropriately, Casaubon's intellectual impotence culminates in a circulatory crisis. No one is surprised when he "suffers a fit in the library," and Lydgate diagnoses exhausted nerves and "fatty degeneration of the heart" (196, 292–93).[22] As his neighbors have always suspected, Casaubon's circulatory system exhibits the same faulty communication evinced by his work and his mind. His "synoptic" project is a dust heap of "shattered mummies, and fragments of a tradition which was itself a mosaic wrought from crushed ruins" (331). Like Eliot's novel, true historicism is a marvel of communication, an intricate, dynamic network that creates organic unity. Just as a nerve net provides the essential links allowing complex organisms to function, good scholarship makes living knowledge possible.

No one can read *Middlemarch* without noticing the affinity of Casaubon's and Lydgate's research projects, both of which explore development and change. Both the philologist and the surgeon seek an "Ur" form, Casaubon's Key to All Mythologies corresponding to Lydgate's "homogeneous origin of all the tissues" (314). Both investigators try to establish connections, but it is here that their projects diverge. Lydgate's quest "to demonstrate the more intimate relations of living structure" differs radically from Casaubon's because any relations that Lydgate discovers must be verified experimentally (*Middlemarch* 102; Cosslett 85–87). Lydgate's interests in development and fevers reveal his basic curiosity about the way the body's structures and functions are interrelated.

Having studied in France, Lydgate has adopted Xavier Bichat's vision of the body as a conglomeration of twenty-one distinct tissues, and he seeks the one, original tissue from which the existing twenty-one have descended.[23] In his textbook *General Anatomy* (1801), Bichat compared these tissues to chemical elements that could give rise to different compounds depending on how they were combined (li–lii). When Eliot describes Lydgate's research, she compares the tissues first to the materials of a house and then to different types of woven cloth, and her similes serve very different purposes. Whereas the invocation of building materials shows how components with different textures can create a unified structure, the alignment with "sarsnet, gauze, net, satin, and velvet" indicates how materials with different textures can share a common origin in the "raw cocoon" (101–2). Taken together, her choices of wood and cloth suggest that a human-made structure—like a written text—can mimic the organic unity of a living body. Eliot could have created an intricate system of web metaphors without any reference to living webs simply by juxtaposing the networks of railways, knowledge, language, and texts. Significantly, however, her incorporation of Bichat's tissues into her metaphorical system associates each of these webs with living networks (Cosslett 94–95). In *Middlemarch*, Bichat's "primary webs or tissues" become a metaphor not just for the social relationships the novel depicts but for the text itself.

While Eliot's comparisons of organic and written texts are finely drawn, they are far from unique. Contemporary scientists were making very similar metaphors. In the 1860s, both Hermann von Helmholtz and Claude Bernard, Europe's leading physiologists, suggested parallels between the ways that signs operated in languages and the ways that the body communicated and organized its many elements. Bernard compared the "simple elements" of living bodies to the letters of the alphabet. In each case, he pointed out, the individual units formed a meaningful structure only through their associations. The meaning of language, like the life of the body, emerged only as a product of the entire system (Shuttleworth 144).

Eliot's comparisons of physiological and language systems are far more complex than those of either scientist, but all three writers imply that well-organized language functions as a body does. For Eliot, one uses language effectively when one uses it to interact with others and form sympathetic ties, promoting the kind of organization that can be found in a living system. As Mrs. Cadwallader tells Dorothea, "we have all got to exert ourselves a little to keep sane, and call things by the same names that other people call them by" (371). Trenchant as she is, she expresses what for Eliot is an essential truth. As cross-points in the social network, people must always heed the thoughts of their neighbors, and they can learn about their feelings only via the sticky threads of language along which all thoughts must move.

Language, however, is an unstable, unpredictable building material. Like Bichat's tissues, with which Eliot juxtaposes it, language has its own life. It is dynamic and changing; it will not be harnessed to the simple tasks of communication and the maintenance of sympathetic links. With dizzying speed, language forges its own connections, but its cross-links lack the logic of those in scientific epistemology. Most conversation in Middlemarch follows a metonymic pattern, so that it is difficult to identify the original object of many discussions. When Chettam discovers Celia and Mrs. Cadwallader speculating about Dorothea's remarriage, he declares the subject "very ill-chosen." Celia protests, "Nobody chose the subject; it all came out of Dodo's cap" (379). In the women's discussion, a reference to Dorothea's unflattering widow's bonnet has led so "naturally" to speculation about its removal that neither one can say how the conversation has moved to the topic she eagerly wanted to discuss. Language, the medium that permits social connections, also allows one to avoid responsibility for one's thoughts.

In Eliot's community of language, one acquires and transmits gossip almost independently of one's will. When Bambridge tells Hawley, "I picked up something else at Bilkley besides your gig horse," the reader at first suspects that he has acquired some unpleasant disease (413). Sure enough, when he releases what he has "picked up"—the news of Bulstrode's duplicity—Bambridge's words move through the social network like a virus.[24] Because "news is often dispersed as thoughtlessly and effectively as that pollen which the bees carry off," one cannot be too careful about one's words (*Middlemarch* 495). By associating gossip with fertilization and disease, Eliot's organic imagery encourages the reader to see language as a living network, as subtle and dynamic as the human body.[25]

Like Bichat's tissues, language must be scrutinized and interpreted. Whether one is reading a text or a body, it is essential to verify one's readings through careful study, but the opportunity for such direct observation does not often arise. Inevitably, hypotheses about tissues and texts are new "tissues of inferences," and like the scientists of her day, Eliot suggests that the best way to interpret is to hypothesize as little as possible (McCormack 49). "Too great readiness at explanation," warns the narrator, "multiplies the sources of mistake" (306). Although the narrator herself speculates endlessly, she warns that "our tongues are little triggers which have usually been pulled before general intentions can be brought to bear" (252). Dissociated from the brain, the tongue is a mere mechanical device, and like a machine, it can cause terrible destruction if one activates it without considering the effects on all those to whom one is connected. In Middlemarch, to speak is to act, and awareness of the web of language in which one lives encourages responsible speech and action. Brooke's statement that "there's no knowing what may be the effect of a vote" resonates powerfully with the narrator's question, "Who shall tell what may be the effect of writing?" (347, 284). Both voting and writing, the narrator suggests, are actions that affect others and have consequences of which one must be aware. The general intentions all speakers must heed are not their own but those of an entire society.

Whenever one speaks or writes, one runs the risk of being misunderstood, and whereas good communication strengthens the social network, misunderstandings can strain or even sever social bonds (Stwertka 184). Unfortunately, the inherent differences between language and the subjects it represents invite such misconceptions.[26] While still an editor, in 1855, Eliot confessed, "no one has better reason than myself to know how difficult it is to produce a true impression by letters, and how likely they are to be misinterpreted" (Haight 189). In the spring of 1869, as an established fiction writer, Eliot warned that "letters are necessarily narrow and fragmentary, and, when one writes on wide subjects, are liable to create more misunderstanding than illumination" (Cross 3: 63) While *Middlemarch* presents most communication as a "good thing," the novel also expresses a deep skepticism about language's ability to represent reality and about people's ability to interpret each other's representations. Just as Dorothea's shortsightedness provides a literal, laughable portrayal of the characters' inability to see, Mrs. Garth's advocacy of correct grammar ("so you should be understood") offers an equally amusing caricature of their struggles to communicate (169). *Middlemarch* abounds with skewed perceptions and misreadings, but they are never the result of myopia or grammatical mistakes. Fred's bold assertion that "all choice of words is slang" seems much closer to Eliot's own position on language, a slippery medium that offers glorious opportunities to communicate and create (67). While the narrator regrets that "we all of us . . . get our thoughts entangled in metaphors," the rich assemblage of figures she offers celebrates the chaotic web of language (57).

Misunderstandings, Eliot implies, originate not in the language network itself but in people's irresponsible use of it. As innocent Dorothea marvels at Casaubon's mind, the narrator reflects, "signs are small measurable things, but interpretations are illimitable" (15). She repeats her warning when describing the townspeople's reading of Lydgate, reflecting that a man "may be puffed and belauded, envied, ridiculed, counted upon as a tool and fallen in love with . . . and yet remain virtually unknown—known merely as a cluster of signs for his neighbors' false suppositions" (96). While the threads of language penetrate everywhere, they will not always be spun into connections based on keen observation and sympathetic understanding.[27]

Networks of Desire

For both Eliot and Lewes, weaving represents the working of consciousness. Not all of the connections the mind makes are valuable, however, and both writers use webs to depict the "wadding" of personal delusions as well as the vital links of productive thought.[28] When Lewes quotes Eliot in *The Physical Basis of Mind,* he draws on her imagery to condemn scientists "spinning an insubstantial universe" of untested theory. For Lewes, "the superstition of the nerve cell" is just such a web, a sticky mesh whose strands fill scientists' heads but lead to no solid structure in the outside world. Like any hypothesis created from personal desires alone, imaginary anatomy is a web spun purely from within.

For Eliot, the spinning of inward consciousness occurs constantly, and when unchecked by meaningful communication with others, it creates egotistical fantasies isolating a mind from the world. In Casaubon, for instance, suspicion and jealousy are "constantly at their weaving work" (290). Because the scholar sees Will and Dorothea only in terms of their effect on him, he can feel only "that proud narrow sensitiveness which has not mass enough to spare for transformation into sympathy, and quivers thread-like in small currents of self-preoccupation" (193). Like the pitiful cobweb of premarital acquaintanceship, Casaubon's net is insubstantial, strung with weak and inadequate material. Sympathetic connections—based on awareness of other people's feelings—make hardy cables, but associations created from selfish, limited perspectives are only the finest threads.

Unhappy Bulstrode, who has been spinning personal fictions for years, has created a tissue of lies so finely woven that he has almost screened off his view of the world. In this hypocritical character, who has acquired his wealth by concealing his knowledge of Will's parentage, "the years had been perpetually spinning [his pleas] into intricate thickness, like masses of spider-web, padding the moral sensibility" (426). Far feebler than a true communications net, such a structure will collapse when forced to incorporate external facts. Just as Lewes predicts the dissolution of "imaginary anatomy," Eliot's narrator remarks, "we are all humiliated by the sudden discovery of a fact which has existed very comfortably and perhaps been staring at us in private while we have been making up our world entirely without it" (228). In science and everyday life, genuine knowledge can be constructed only by observing and by comparing one's experiences to those of others.

Eliot's most powerful representation of egotistical perception is her fascinating pier-glass metaphor:

> Your pier-glass or extensive surface of polished steel made to be rubbed by a housemaid, will be minutely and multitudinously scratched in all directions; but place now against it a lighted candle as a center of illumination, and lo! the scratches will seem to arrange themselves in a fine series of concentric circles round that little sun. It is demonstrable that the scratches are going everywhere impartially, and it is only your candle which produces the flattering illusion of a concentric arrangement, its light falling with an exclusive optical selection. These things are a parable. The scratches are events, and the candle is the egoism of any person now absent. (182)

Strictly speaking, this illusory pattern of scratches is not a web. It consists of concentric circles, whereas a web depends on intersecting lines. Arguing that the web image has its origin in geometry, Selma Brody believes that it reflects Eliot's passion for mathematics. Because a web has geometrical symmetry, she points out, "every intersection is the equivalent of any other, yet from the viewpoint of any one intersection the entire lattice seems centered on that particular point" ("Physics in *Middlemarch*" 45). As a symbolic image, this parable bears a close affinity to Eliot's networks of knowledge, language, and personal fantasy. Like the web metaphors, the pier-glass parable urges readers to rethink their perceptions and strive toward a more global vision by comparing their views with those of others.

In their references to innumerable fine lines, both the web and the pier-glass metaphors suggest Faraday's lines of force.[29] In his *Fragments of Science for Unscientific People,* John Tyndall asks his readers to "look with your mind's eye at the play of forces around a magnet." When a magnet is placed under a paper with fine pieces of iron wire scattered over it, he writes, the fragments will "embrace the magnet in a series of beautiful

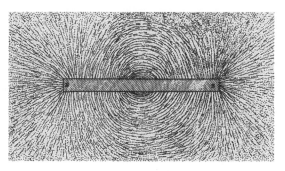

Michael Faraday's (*above*) and John Tyn-
dall's (*left*) representations of the lines of
force around magnets. (Michael Faraday,
Experimental Researches in Electricity, 3
vols., London: Richard Taylor and William
Francis, 1839–55, vol. 3, plate 4, fig.2, and
John Tyndall, *Fragments of Science* [New
York: Appleton], 1: 362.)

curves." As we view these lovely patterns of fine lines, "we soon recognize a
brotherhood between the larger phenomena of Nature and the minute
effects which we have observed in our private chambers" (1: 363–64). The
lines of force that Tyndall describes, however, are a scientific "fiction"
woven to explain the action of magnetic forces. They do not exist in space,
although the magnet does arrange the metal fragments into actual lines. In
Eliot's metaphor, the light of egotistical vision creates a pattern that seems
equally real. The magnet and the egotistical eye are not perfectly analogous,
because the magnet exerts an actual physical force whereas the eye merely
observes, creating structures only in the mind. Both, however, help to produce the patterns that the mind perceives. Eliot's pier glass, like her webs,
depicts the play of forces among which individuals live and move.[30]

In the society Eliot depicts, all individuals are magnets of sorts, and the
forces they exert are their desires. These longings make up yet another web,
created by the interference patterns of their many intersecting lines.[31] Like
knowledge, Eliot suggests, desire is a woven text. Unbeknownst to Lydgate,

Rosamond has "woven a little future" with him in it (80). And Peter Featherstone's relatives, discovering the existence of multiple wills, fear that "there might be such an interlacement of poor Peter's former and latter intentions as to create endless 'lawing'" (230). To maintain one's identity in such an environment and to exert a valuable and meaningful influence upon others, one must be aware of the way other people's forces influence one in turn. To do "great work for the world," one must feel the desires of one's neighbors and see one's own wishes from the perspective of the social body.

Conclusion: The Sympathetic Network

Like the physicists and physiologists of her day, Eliot used networks to represent the transmission of information, the construction of knowledge, the play of language, and the operation of desire. In their daily lives, she and the scientists encountered the same images of communications networks, and in all of their writing, web metaphors emerge in passages promoting circulation and connectedness. It is hardly surprising that in a culture building communications nets, writers should see them in the body, in communities, and in the structure of knowledge itself. Also like the web metaphors of some scientists, Eliot's imagery serves a moral system. Like Lewes's, Eliot's images of interconnected points remind readers that life and mind have meaning only in relation to entire organisms. While each individual cell contributes to the being of which it is a part, it is never independent of that organism. Sympathy, for Eliot and Lewes, is as essential to society as it is to the body.

The word "sympathy" recurs throughout Eliot's writing, arising as frequently in her essays and letters as in her fiction. In "The Natural History of German Life," she asserts that "the greatest benefit we owe to the artist . . . is the extension of our sympathies" (Eliot, "National History," 582). For Eliot, sympathy means not just shared feelings but any sort of communication leading to genuine understanding and mutual agreement. She and Lewes, for instance, had a "lively sympathy" on the subject of proofreading (*George Eliot Letters* 5: 98). In the eighteenth century, sympathetic connections were believed to keep bodily organs informed of each other's conditions. Eliot's extensive use of the term reveals the many parallels she saw between organic and social relationships.

For Eliot, morality began with an awareness of how one's life affected the lives of others. In *Middlemarch*, the narrator comments ironically that "we are most of us brought up in the notion that the highest motive for not doing a wrong is something irrespective of the beings who would suffer the wrong" (172). There is, she implies, an alternative motive: an honest consideration of how one's action might hurt other people. When describing Bulstrode's moral decay, Eliot implies that human feelings, not accepted ideas, should govern our behavior. "There is no general doctrine," proclaims the narrator, "which is not capable of eating out our morality if unchecked by the deep-seated habit of direct fellow-feeling with individual fellow-men" (428). Like the scientific methods she studied, Eliot's morality is inductive, a body of hypotheses always receptive to new inputs. "A moral system," she writes, "to govern society, must accommodate itself to common characters and mingled motives" (Pratt xli). When she depicts a web of social relationships, her thinking reflects the scientists' not because she directly incorporated their ideas but because they all subscribed to many of the same codes of ethics.

Like one's body, Eliot believes, one's moral sense develops in a dynamic, ongoing process. When the narrator asserts that "we are all of us born in moral stupidity, taking the world as an udder to feed our supreme selves," she implies that this lamentable state is merely a beginning, a phase that can (though will not necessarily) be outgrown (146). Sympathy, the novel implies, is not an endowment. Sympathy grows, and one must always encourage its growth. To progress beyond this stage, one needs "the remedial influences of pure, natural human relations" (Haight 341). The more one interacts with others, Eliot believes, the more one will learn about the real meaning of one's actions.

For Eliot, one learns to form sympathetic connections by developing one's vision. To a certain degree, she asks readers to suppress their selfish perspectives as scientists supposedly do, in order to achieve a more "objective" view. This initial distancing, she hopes, will make people feel their "participation in a wider structure" (Cosslett 74). Eliot knows such an assessment of scientific vision is naive, of course; one need only read Lewes's discussions of nerve cells to see how deeply a scientist can be involved in his theories. Like Tyndall's, the vision that Eliot demands is both objective and subjective. She decries egotistical perception but demands sympathetic identification with others (Miller, "Optic and Semiotic" 136). Lewes and Tyndall pointed out the central role of imagination in science, and Eliot asks that her readers use their imaginations to envision their position in a complex social organism.

Dorothea's visual and moral development in *Middlemarch* illustrates what Eliot saw as all people's potential to learn and grow. Her heroine begins with a "fanaticism of sympathy," but her sensitivity proves valueless until she understands her relationships with others (*Middlemarch* 152). Eliot liked to laugh at herself, and Dorothea's childhood belief in "the gratitude of wasps and the honorable susceptibility of sparrows" shows the author's awareness that even respect for one's fellow creatures can be taken to unrealistic extremes (150). Dorothea is flawed, above all in her vision, but her modest, unhistorical success offers hope to all who read her history. Her final feeling of connection to "that involuntary, palpitating life" is one that all people can achieve if they learn to see themselves as intersections in a network rather than as the foci of concentric circles (544).

Like Lewes's nerve nets, Eliot's web metaphors convey the organic structure of society (Shuttleworth 3–5). Her images repeatedly suggest that the best vision, the best knowledge, and the best action are all based on the awareness of connections—on consciousness of one's position in a web. As a cross-point in a network, a human being can both influence and be influenced. Eliot agrees with Lewes that "a human being . . . is a very wonderful whole, the slow creation of long interchanging influences" (*Middlemarch* 282). Like Ramón y Cajal, she celebrates the open, dynamic nature of the human mind. Her conclusion, that "there is no creature whose inward being is so strong that it is not greatly determined by what lies outside it," is nowhere near as pessimistic as it sounds (577). While it promises the thwarting of much human desire, it does not preclude happiness. Unhappiness, she suggests, comes from spinning desires without considering one's position in the social web.

Eliot's praise for the railroad and telegraph networks which were "making self-interest a duct for sympathy" reflects her profound belief that communication fosters moral growth. Though built for profit, these networks promised to connect communities just as the nervous system connected all regions of the body, and the sympathy they made possible was a cause for celebration.

Chapter 4
The Language of the Wires

Like George Eliot, the scientists and engineers who designed nineteenth-century communications networks viewed their growing webs as organic structures. A decade before DuBois-Reymond performed his electrophysiological studies, Samuel Morse wrote, "it [will] not be long ere the whole surface of this country [is] channeled for those *nerves* which are to diffuse, with the speed of thought, a knowledge of all that is occurring throughout the land" (Morse 2: 85, original emphasis).[1] Years later, describing the first attempt to lay a transatlantic cable, Morse's son and biographer echoed his father's metaphor: "thus ended the first attempt to unite the Old World with the New by means of an electric nerve" (Morse 2: 382). Just as Helmholtz saw nerves as telegraphs, Morse and other designers of telegraph networks saw their wires as nerves. Inspired in part by electrophysiology, they set out to build communications systems that matched the body's own ability to transmit messages.

From the late 1840s onward, the feeling of unity created by telegraph lines inspired both scientific and lay writers to compare technological and organic communications systems. With little effort, they incorporated telegraph lines into the long-standing metaphorical system describing society as a living body. Because organic webs—not just spiders' webs but the webs of living tissue—resembled telegraph networks physically as well as functionally, they became the perfect metaphorical vehicle for these new communications systems. Consequently, writers from many fields described the growing networks in organic terms. In *The Story of the Telegraph* (1858), published just after the completion of the first transatlantic cable, Charles Briggs and Augustus Maverick asked their readers to imagine a world "belted with the electric current, palpitating with human thoughts and emotions" (12). Henry Field, who financed the transatlantic line, called it "a living, fleshy bond between severed portions of the human family," and after it was successfully laid in 1866, people toasted it as "the nerve of international life" (Standage 104, 91). Echoing George Eliot in an article on the "moral influence of the telegraph," a writer for *Scientific American* proclaimed that "the touch of the telegraph key weld[s] human sympathy and ma[kes] possible its manifestation in a common universal, simultaneous heart throb" (Standage 162). As a metaphorical vehicle, the organic web allowed writers to carry organic and technological communications systems into new realms of meaning and encouraged readers to consider the implications of their similarities. The power to communicate promised to unite societies with new, living bonds.

This understanding of the telegraph as a nerve network affected engineers' decisions about how to construct it. Like Galvani, many scientists who experimented with electricity incorporated living receivers and conductors directly into their circuits. In 1767, Johann Georg Sulzer (1720–79) discovered that when one applied two different metals to the tongue, one experienced a strange tingling and an odd taste (Sabine 14). Many eighteenth-century investigators, including Alessandro Volta, used living tongues as "receivers." Although Volta did not know of Sulzer's work, Volta developed his battery as a result of his own electrical experiments with organic systems (Pera 182).

As organic receiving devices, the body's sensory organs suggested ways to detect, translate, and record a great variety of signals. The fact that so many communications engineers made practical use of the body's own "wires" suggests that the similarity between organic and technological networks is more than an analogy. Frustrated in their attempts to develop technological receivers as sensitive as the body's, some scientists incorporated organic receivers into their telegraphic systems. In the first decade of the nineteenth century, Vorselmann de Heer proposed an "electro-physiological telegraph," in which the ten fingers of a willing subject served as the receiving device. The problem with his apparatus, he complained, was that one had to keep altering the voltage according to the receiver's sensibility (Prescott 56). Werner von Siemens had workmen stick their hands into chambers of water, relying on the mild shocks they got to detect faults in his cables' insulation (Siemens 73). In the mid-1860s, while redesigning the transatlantic cable, physicist William Thomson applied the telegraph wires to his own tongue and found that he could indeed "taste" differences among signals.

Like a human tongue, a telegraph key could speak as well as taste. Recounting William Cooke's first anxious demonstration of the British needle telegraph to railway engineer George Stephenson, Latimer Clark described Cooke's receiver as a "trembling tongue of steel" (88). The problem with the telegraph in its earliest incarnations, wrote Briggs and Maverick, was that there was no way of "making it speak" or "reporting and preserving its utterances" (24). On a Morse instrument, an experienced operator could "read" incoming messages just by "listening to the clicking of the apparatus" (Standage 65). Reflecting that "the clattering tongue of brass seems alive as I listen and hear the signals pass," an unknown telegrapher-poet recognized as fully as Volta, Thomson, or Helmholtz the affinity among the tongue, the ears, and the telegraph key as sophisticated transmitting and receiving devices (Johnston 111).[2] Almost all of the scientists who made the greatest contributions to telegraphy had carefully studied the body's own sensory "devices." During the 1830s, the electromagnetic telegraph was perfected by an anatomist, an artist, several physicians, and a group of optical and acoustics experts.

Early Attempts to Communicate over the Wires

Since the mid-nineteenth century, the term "telegraph" has implied an electromagnetic communications system, but originally it denoted any system of signals used to "transmit intelligence." In the eighteenth century, telegraphy encompassed every sort of visual signaling, including fires, beacons, and semaphore systems. By the 1790s, these systems were growing increasingly sophisticated, and the 1794 French Directory declared that "the new invented telegraphic language of signals is an artful contrivance to transmit thoughts in a peculiar language from one distance to another" (Francis 2: 272). When nineteenth-century writers described these earlier systems, they used the same words that they did for the later electrical telegraph. Like nerve nets, these systems either "conveyed," "transmitted," or "communicated" intelligence (Ronalds 1; Morse 2: 183; Sabine 6).

Since prehistoric times, groups of people had used drums and signal fires to send messages over a distance. Once the telescope became available in the seventeenth century, governments and businessmen began deploying chains of messenger-observers along hilltops within visual range. As long as these sentries remained alert and the daylight and weather held out, they could relay information from city to city infinitely faster than messengers on horseback (*New Encylopedia Britannica* 28: 474). In France, in the early 1790s, Claude Chappe developed a device on which rotating arms mounted on a bar could assume ninety-eight different positions, each corresponding to a symbol detectable by an observer in a nearby tower. Chappe's system proved so successful that by the 1830s there were lines of telegraph towers all over Europe, a "mechanical Internet of whirling arms and blinking shutters" (Standage 16). In 1832, a group of Wall Street brokers established such a chain to transmit market prices to Philadelphia businessmen. Placing

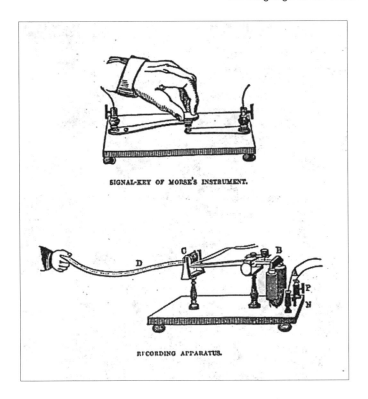

Above, Charles F. Briggs's and Augustus Maverick's drawings of a Morse telegraph apparatus, including the hand of its operator. Compare to Luigi Galvani's circuits on page 17. (Briggs and Maverick drawings from Charles F. Briggs and Augustus Maverick, *The Story of the Telegraph and a History of the Great Atlantic Cable* [New York: Rudd and Carleton, 1858], 29.)

observers with telescopes on hills about six miles apart, they were able to transmit messages at a rate of 360 miles per hour (Asmann 5).

The efficiency of visual telegraphs—on sunny days—permitted them to survive well into the nineteenth century, but in foggy Northern Europe their unreliability became the butt of numerous jokes. In 1846 the British journal *Punch* offered readers a story that while probably fictional, drew its humor from popular experience:

Last week, the French telegraph made the following announcement—"Abd-el-Kader has been taken"—but it was mentioned that a fog enveloped the remainder of the sentence in obscurity. The funds, however, rose tremendously. The following day, the sentence being completed, the intelligence ran thus: "Abd-el-Kader has been taken with a dreadful cold in his head." The funds fell, but the coup had been sufficiently successful for those who made the telegraph play into the hands of their agents at the Bourse. A fog in Paris is frequently a great windfall. ("Tricks of the Telegraph" 35)

Until the mid-1840s, however, the visual telegraph remained the dominant system of rapid, long-distance communications, for the electrical telegraph appeared even less dependable. The U.S. Congress had a visual telegraph in mind when it requested proposals for a national system in 1837, but Samuel Morse urged Congress to use an electrical one instead.

In the latter half of the eighteenth century, many scientists had suggested using electricity to transmit information. In a letter to *The Scot's Magazine* in 1753, an inventive thinker identified only as "C.M." suggested "an expeditious method of conveying intelligence," a system of twenty-six wires, activity in any of which would cause an attached ball to tremble at the receiving end (Kieve 14; Sauer 2; Standage 17–18).[3] Between 1753 and 1837, scientists all over Europe developed more than sixty different models for a device that could convey intelligence by means of electrical signals (Standage 18).

No one "invented" the telegraph, just as no one invented the railway. Instead, both emerged gradually as a variety of individuals developed new insights based on their studies of organic and electromagnetic systems (Sabine 40). Volta, whose battery proved essential for the telegraph, had thought of sending signals along wires in 1777 (*New Encyclopedia Britannica* 28: 474). In 1795, Italian physicist Tiberio Cavallo (1749–1809) suggested "transmit[ting] letters and numerals by combinations of sparks and pauses," and Spaniard Francisco Salva connected Madrid and Aranjuez with a system using combinations of signals along several wires to represent letters of the alphabet (Kieve 15; Prescott 59–60; Sabine 10). Seeking a new way to reveal the presence of electricity in a circuit, the Bavarian scientist S. T. Sömmering (1755–1830) built a device in which hydrogen bubbles, released when water decomposed, caused an alarm bell to ring (Kieve 14). When Hans Christian Oersted discovered in 1820 that electric currents could deflect magnetic needles, French physicist André Marie Ampère took advantage of the idea to build an electromagnetic telegraph (Briggs and Maverick 19; Sauer 8). The most successful and best-known model of the telegraph prior to the 1830s was probably that of British inventor Francis Ronalds (1788–1873), who adapted the wheels of a clock to send and receive electrical signals in coded form (Ronalds 4–9; Sabine 10–12). Seeing the possibilities of a telegraphic network even in 1823, Ronalds urged, "let us have *electrical conversazione offices,* communicating with each other all over the kingdom, *if we can* . . . give me *materiel* enough, and I will electrify the world!" (2, 21, original emphasis).

Alexander von Humboldt, who began his career by experimenting with animal electricity, proved as vital a link among scientists designing the telegraph as among those exploring the nervous system. In 1828 he invited the eminent mathematician Karl Friedrich Gauss (1777–1855) to a meeting of natural scientists in Berlin, hoping to enlist Gauss in his scheme to build a worldwide network of mathematical observatories (Gillispie, *Dictionary* 5: 304–5). At this meeting, Gauss met Wilhelm Weber (1804–91), a young acoustics expert who had been introduced to science by E. F. Chladni and had come to the conference to deliver a talk on organ pipes (Gillispie, *Dictionary* 14: 204). Deeply interested in biomechanics as well as in acoustics and optics, Weber frequently collaborated with his younger brother, an anatomist. In the 1830s, while Weber was working with Gauss, the two brothers were analyzing the physics of human bodily motion. In their fruitful collaboration between 1831 and 1837, Gauss and Weber studied the earth's magnetic properties and developed a new system of units for measuring the strength of magnetic fields. In 1833, they constructed a highly successful electromagnetic telegraph—an uninsulated copper wire strung over the rooftops of Göttingen. Their goal was not to improve worldwide communications but merely to transmit information more efficiently between their astronomical observatory and physics lab. Gauss, always interested in numbers theory, saw quickly that "a limited number of signs [was] all that [was] required for the transmission of messages" (Briggs and Maverick 20). When politics ended their collaboration in 1837, Weber continued to study the behavior of electricity. He and his coworker Rudolph Kohlrausch (1809–58) later showed how slowly Helmholtz's nerve impulses were traveling when they accurately measured the velocity of electricity in 1856.

Weber, however, was not the first physicist to measure the velocity of electricity. In 1834, using a rotating mirror to detect sparks emitted by an electrical circuit, British experimentalist Charles Wheatstone had calculated the speed to be over 250,000 miles per second, about 1.3 times the speed of light (Prescott 63; Gillispie, *Dictionary* 14: 290). Like Weber, Wheatstone had begun his career not as a physicist but as an acoustics expert. Coming from a family of flute makers, he spent fifteen years studying the properties of sound before he ever turned to electricity or telegraphy. Throughout his career, he maintained deep interests in optics, acoustics, and cryptography. He was forever fascinated by the body's ability to detect and process information of all kinds. Wheatstone observed and participated in mesmeric experiments, in 1838 serving on a Royal Society committee to investigate mesmeric phenomena (Winter, *Mesmerized* 95, 49–50). With Michael Faraday, he studied the electricity generated by electric fish, but his greatest interest was always the human body's sensory modalities. The stereoscope he invented made viewers feel as though they were confronting three-dimensional scenes, and his kaleidophone translated sounds into visual images (Winter, *Mesmerized* 39). As a means of representing sound waves visually, Chladni's acoustical figures particularly intrigued Wheatstone. While he is best known for perfecting the needle telegraph jointly with William Cooke, he was already wondering, decades before Alexander Graham Bell, how one might transmit speech over long distances (Gillispie, *Dictionary* 14: 289).

Wheatstone's partner, Cooke, the son of an anatomy professor, began his career building wax anatomical models for medical students. Cooke's father, to whom he always referred as "the doctor," had been a close friend of Francis Ronalds and had helped him perform some of his telegraphic experiments. It is possible that as a boy, young William may have seen an electrical telegraph in action (Clark 62; Kieve 18). During 1833–34, Cooke studied anatomy and physiology in Paris, where he "practiced with great ardor the pursuit of modeling anatomical dissections in colored wax—an art in which he eventually acquired great skill" (Clark 62). As a young man, Cooke heard again about electrical telegraphs while performing anatomical dissections in Heidelberg (Standage 30–31). In the 1820s, following Oersted's discovery, the Russian diplomat Pavel Lvovitch Schilling had developed a galvanometer-like telegraph in which five magnetic needles pointed to letters in response to a flow of current. While Schilling died before his instrument became widely known, he made several copies, one of which German naturalist Georg Muncke (1772–1847) demonstrated to Cooke in a lecture on electromagnetism (Kieve 17; Standage 30–31). When Cooke saw Schilling's telegraph, he vowed to work full-time designing an electrical communications system, but he never completely abandoned his earlier scientific training. After years of molding nerve tracts out of wax, he used the same word, "model," for his telegraphic apparatus as for his earlier creations.[4] His descriptions of his efforts suggest that he approached organic and technological communications systems the same way.

In 1837, the race to develop an electrical telegraph reached its final, most intense stages. In June, Cooke and Wheatstone obtained a British patent for their instrument, an improved version of Schilling's needle telegraph (Sauer 8; Kieve 17–18). In July, German optician Karl August Steinheil (1801–70) opened a telegraph line between Munich and Bogenhausen whose receiver involved a needle, a clockwork mechanism, and a system of markings to represent letters (Briggs and Maverick 24–25). Steinheil, who between 1849 and 1852 would construct the Austrian telegraph network, improved communications technology by showing that the earth itself could take the place of a ground wire, completing the telegraphic circuit (Prescott 54). He was equally well known, however, for the optical workshop he founded and the telescopes he designed (Gillispie, *Dictionary* 13: 22).

While Steinheil did not challenge Cooke and Wheatstone's claim on the needle telegraph, a British surgeon, Edward Davy, contested their petition for a patent. In the spring of 1837, Davy was conducting experiments using a system he had rigged in Regent's Park. It is quite possible that the doctor's telegraph was working before Wheatstone and Cooke's, but his instrument was judged to be inferior, and he lost the suit. William Alexander, a Scottish inventor, also opposed Cooke and Wheatstone's patent based on an alternate system that he proposed in July 1837 (Kieve 23–24). The most serious competitor of the British anatomist and the electrician, however, was an American painter, a designer who had been working independently since 1832 on a device that would use electricity to transmit intelligence instantaneously.

In October 1832, following a shipboard discussion of electricity and magnetism, Samuel F. B. Morse suddenly "saw" how electricity might be used for communication (Morse 2: 5–7).[5] At that time, the young painter was returning from Paris, where he had been studying artistic techniques. Hearing about European scientists' new experiments, Morse was struck by "the fact that electricity passed instantaneously through any known length of wire, and that its presence could be observed at any part of the line by breaking the circuit." It occurred to him that "if the presence of electricity can be made visible in any part of the circuit, [there is] no reason why intelligence may not be transmitted instantaneously by electricity" (Morse 2: 6). Like Helmholtz, Morse liked to explore ideas visually, and he immediately sketched out a potential circuit in which "breaks" corresponded to letters of the alphabet. In his initial diagram, he included a code of dots and dashes.

Five years later, in a letter to Congress offering his system for national use, Morse claimed that he had originally been inspired by Benjamin Franklin's (1706–90) transmission of electricity across the Schuylkill River in 1748. But James Fenimore Cooper (1789–1851), who spent the winter of 1831–32 in Paris with Morse, remembered him speaking of "using the electric spark by way of a telegraph" in their conversations at that time (Morse 1: 419). While Morse was studying painting in Paris that year, he met a number of European scientists, including Alexander von Humboldt (Morse 2: 365). Although Humboldt had performed his electrophysiological studies decades earlier and by 1832 was principally a geographer, his transmission

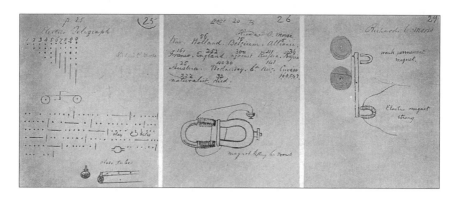

Samuel Morse's 1832 sketch of a circuit for "transmitting intelligence" with electrical signals, including a code of dots and dashes. (From *Samuel F. B. Morse: His Letters and Journals,* ed. Edward Lind Morse [Boston: Houghton Mifflin, 1914], 6.)

of Helmholtz's paper to the French Academy of Sciences in 1850 indicates his ongoing interest in the relationship between electricity and nerve impulses. Morse's interaction with Humboldt and the European scientific community of the early 1830s suggests that he was thinking about animal as well as common electricity.

By the 1830s, despite scientists' skepticism about Galvani's results, everyday language was encouraging an identification of organic and electromagnetic forces that scientists hesitated to make in their reports. Popular interest in animal electricity merged with a growing fascination with mesmerism and "animal magnetism." In 1842–43, during congressional debates about whether to finance an experimental telegraph line based on Morse's apparatus, opponents of the bill used this ambiguity as a way to cast doubt on Morse's model.[6] Representative Cave Johnson (1793–1866) demanded that half the money be set aside for mesmeric experiments, stating that he "did not wish to see the science of mesmerism neglected and overlooked." Representative Stanly replied that he would be delighted to oblige as long as Johnson would serve as subject in these experiments, and the Chair decided that "it would require a scientific analysis to determine how far the magnetism of mesmerism was analogous to that to be employed in telegraphs" (Morse 2: 194–95). A century and a half later, it is difficult to interpret Stanly's and Johnson's exchange. On the one hand, the congressmen—who hardly seem serious—appear to be associating the electromagnetic telegraph with animal magnetism in order to disparage it. As Alison Winter has shown, mesmerism was often used in the 1840s as a way to debate larger social issues like communication (*Mesmerized* 267). On the other hand, the congressmen may genuinely have seen animal magnetism and electromagnetism as identical, expressing what Winter calls "a loosely connected set of inferences based on common beliefs about physical forces" (*Mesmerized* 120). The *Congressional Globe,* which covered the debates, entitled its article "Electro and Animal Magnetism" (Morse 2: 193). The company that built the first Morse lines was called the "Magnetic," and when the public heard this word, they envisioned both Faraday's lines of force and strong-willed mesmerists who manipulated their subjects.

Throughout the 1840s and 1850s, the American and European public—including physiologists like Helmholtz and DuBois-Reymond—witnessed the explosive growth of telegraph networks (Standage 57–59). In 1849, the United States had 11,000 miles of telegraph lines; in 1850, Great Britain had a mere 2,200. By 1854, the United States had 42,000 miles; Great Britain 40,000; France 18,000; and Prussia about 4,800. By 1867, the United States had 90,000 miles; Great Britain 80,000; France 70,000; and Prussia 45,000 (Sauer 9–11). In 1860, George Prescott reported that U.S. telegraph lines were "creeping over the Rocky Mountains," and by 1861 the first transcontinental line was complete (Prescott v–vi). A transatlantic cable connected Europe and America fleetingly in 1858, then permanently in 1866. As one writer had predicted in 1848, the world was becoming "covered with network like a spider's web" (Standage 58).

This new network did not spring up of its own account. As all of the telegraph's designers conceded, its lines followed those laid out by the railways in the late 1830s and early 1840s, the circulatory system suggesting pathways along which information should flow (Sauer 9; Kieve 13).[7] To develop fully, however, the railways needed the telegraph. Railway expert Max Maria von Weber proposed that:

> as the muscle of a human body without the nerve flashing through it would be a mere lifeless hunk of flesh, so would the flying muscles that Watt's and Stephenson's inventions have lent to humanity be only half as capable of winging their way, if they were not animated by the guiding thought imperiously flashing through the nerves of the telegraph wires. (Schivelbusch 30)

The railroads depended heavily upon signaling because knowing the positions of trains not only increased efficiency; it became a matter of life and death. Often these "nerves" ran parallel to railroad tracks, with railway stations doubling as telegraph offices. In Prussia, the wires were buried in rail-

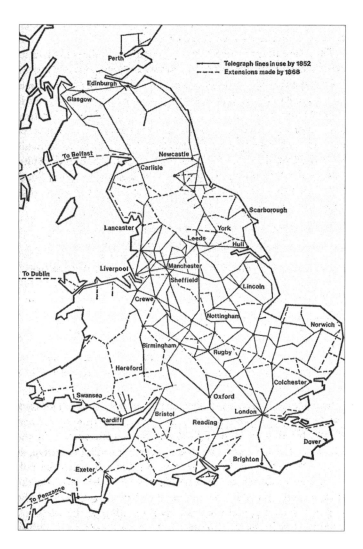

The growth of England's telegraph network between 1852 and 1868. (From Jeffrey Kieve, *The Electric Telegraph: A Social and Economic History* [Newton Abbot: David and Charles, 1973], 75.)

way embankments under a few feet of loose soil. In the mid-1830s, Karl Steinheil had even considered using the iron rails themselves as conductors. Werner von Siemens, who designed Prussia's communications net in the early 1850s, believed that telegraphy was "undergoing rapid development" at that time "in consequence of the introduction of the railways" (34). In England, the success of the telegraph was assured when engineer George Stephenson told his directors, "Let Mr. Cooke have everything he may require" (Cooke 36).

In the United States, Congress finally yielded to Morse's arguments and in 1843 approved thirty thousand dollars for an experimental telegraph line from Baltimore to Washington. In 1844, the first public telegraph became operational. While the government seriously doubted the value of this new communications system, the telegraph eventually proved useful. Besides allowing officials in Baltimore and Washington to play real-time chess games, it helped the police to apprehend criminals and the army to locate deserters (Asmann 16; Standage 50, 53).

It would be many years, though, before the general public appreciated their new "nervous system."[8] In its early years, the telegraph was valued mainly by particular communities—railway engineers, businessmen, and later, journalists—who relied on the rapid transmission of information. Besides railway directors, the most important backers of the telegraph were traders and brokers who benefited from the rapid dissemination of market prices. Throughout the late 1840s and the 1850s, they gladly offered capital for the construction of new lines. By 1870, William Orton, president of Western Union, could boast that the telegraph was "the nervous system of the commercial system" (Standage 170). Journalists initially saw the telegraph as an "ominous development," because it ended the era of foreign correspondents and the race to get the scoop on foreign news (Standage 149). Once newspapers grew accustomed to the new "lightning communication," however, they quickly came to depend upon it for news "flashes" from distant cities. Gradually, all those who could afford to do so began to experience the value of the new instantaneous communication.

As the new telegraphic webs became functional in the late 1840s, people could know—within limits—what was occurring in distant cities at any particular moment (Asmann 16). By eliminating communications delays people had always taken for granted, the telegraph altered people's concepts of space and time. Demonstrating his system to a congressional committee in 1838, Samuel Morse boasted that "space [is] annihilated" (Asmann 11). Because the telegraph's filaments unified societies to a degree never before known, the distance between cities became much less meaningful. People—operators, at least—could talk to each other in "real time."

Like nerves, wires had their vulnerabilities. In the nineteenth century, the greatest difficulty faced by engineers trying to transmit electricity over significant distances closely paralleled one of the nervous system's greatest evolutionary challenges: the problem of insulation. To transmit electrical signals efficiently, any communications system, organic or technological, had to protect its lines with some nonconducting material. Cooke, Morse, and other designers of the 1830s found that their circuits leaked, and the Voltaic batteries that drove them could not maintain enough current in the wires to make communication feasible. Sending electrical signals over significant distances was not possible at all until American physicist Joseph Henry (1797–1878) developed the relay system, dividing very long circuits into many smaller ones, each of which boosted signals (Briggs and Maverick 24).

In marine and underground lines, insulation was critical. The transatlantic cable of 1858 failed because its designer, surgeon Edward Whitehouse, tried to overcome the diminution of signals by using extremely high voltages (Mullaly 45; Standage 78). Struggling to protect elevated, underground, and eventually submarine cables against frequent breaks, engineers tried encasing their wires in shellacked cloth; resins; lead and iron pipes; and finally gutta-percha, the gum of a Southeast Asian tree. Werner von Siemens, under pressure in 1850 to construct the underground Prussian network as quickly as possible, systematically studied the properties of electricity in submerged cables. Siemens deserves far more credit than he has received for selecting gutta-percha as the best possible insulator (Siemens 51, 88; Standage 70). Following his early attempts to detect leaks with his workmen's fingers, Siemens greatly improved the technology for locating faults in the insulation that could lead to breakdowns in the system. Breaks in the wires were described as "injuries," and the workers who maintained lines identified themselves with medical doctors. Anticipating breaks in his lines even in 1823, Francis Ronalds pondered how British engineers might "discover the seat of the disorder, the exact point or points which have sustained injury" (18).

In 1862, a writer for *The Cornhill Magazine*—a British journal for educated lay readers—relied on people's knowledge of telegraph cables to explain the structure of the nerves. Considering the enormous publicity received by the ill-fated 1858 Atlantic cable, it was reasonable to presume such a familiarity. According to this writer:

> the fibers which constitute the chief mass of the nervous system are simple in their structure . . . and present a very curious analogy to a telegraphic wire. Like the latter, each nervous fiber consists of a small central thread . . . surrounded by a layer of a different substance. . . . The white substance acts the

part of the gutta percha round the electric wire, as an insulating medium for the currents which travel along the central portion. ("What are the Nerves?" 157)

To the public, as to the engineers, insulation in the telegraph network seemed to work according to the same principles as that used in the body.

Despite the obvious similarities between the natural and artificial "nervous systems," the body's communications network differed from telegraph networks in several key respects. The nervous system surpassed technological systems in its capacity to remember, and those who built the new communications systems bore in mind not just the sensory organs' abilities to interpret signals but the brain's capacity to process information. From the first, designers like Morse had puzzled over how to create permanent records of the signals they transmitted. Referring to his telegraphic transmissions as "signs," Morse struggled for years to devise an efficient code for sending them. Designers of semaphore systems had shown that a good code compressed information, transmitting it in compact pellets that could be expanded and interpreted at the receiving end. Seeing immediately that a code was an essential aspect of his mechanism, Morse had provided one in his very first sketch, and he spent five years perfecting it.[9] In all probability, it was his invention of the code that led so many to credit him with inventing telegraphic communication as a general concept.

Envisioning a magnetic receiver that would transcribe dots and dashes onto paper, Morse had also included a transcribing device in his sketch of 1832. In his 1837 letter to Congress, he specified that his telegraph was superior to semaphore systems because "the record of intelligence is made in a permanent manner" (Asmann 312). Applying for a British patent in 1838, Morse argued that his telegraph differed significantly from Cooke and Wheatstone's needle model because of its ability to "imprint permanent signs at a distance" (Morse 2: 92). This capacity to produce direct, permanent transcripts may well have made Morse's telegraph more attractive than the early British system and led to its adoption throughout the world. From the beginning, it had taken representation into account, and it mimicked—though it could not equal—the biological function of memory.

The Other End of the Wire: Telegraphic Fiction in 1877

Reflecting on the advantages of Morse's telegraph in 1847, Werner von Siemens praised "the simplicity of Morse's apparatus, the relative facility of acquiring the alphabet, and the pride which fills everyone who has learnt

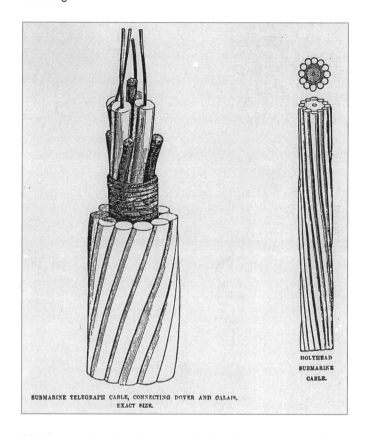

SUBMARINE TELEGRAPH CABLE, CONNECTING DOVER AND CALAIS. EXACT SIZE.

HOLYHEAD SUBMARINE CABLE.

Charles F. Briggs's and Augustus Maverick's representations of undersea cables and their layers of insulation. (From Charles F. Briggs and Augustus Maverick, *The Story of the Telegraph and a History of the Great Atlantic Cable* [New York: Rudd and Carleton, 1858], 34, 35.)

how to use it, and which causes him to become an apostle of the system" (82). A gifted technician, Siemens could see that the Morse apparatus worked so well because it exploited the operators' pleasure in their own manual dexterity. The genius of the new communications device was that it incorporated the hands, ears, nerves, and brain of its operator, who merged with it to form a doubly empowered device for transmitting and receiving information.

But the human elements of these new communications devices challenged designers' claims about their benefits. Morse and the promoters of the transatlantic cable had envisioned new, living bonds that would promote

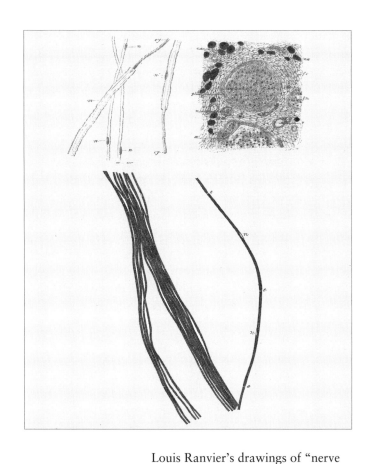

Louis Ranvier's drawings of "nerve tubes" (*Nervenröhren*). *Above left,* an adult rabbit's sciatic nerve, whose torn end reveals the emerging axon. *Above right,* cross section of a collateral nerve in the human finger, showing the "layered sheath" (*gl*). *Left,* a rabbit's sciatic nerve drawn out in water following osmium fixation. (From Louis Ranvier, *Technisches Lehrbuch der Histologie,* trans. W. Nicati and H. von Wyss [Leipzig: Vogel, 1877].)

Schematic representation of a nerve "cable." (From "What Are the Nerves?" *Cornhill Magazine* 2 (1860): 157.)

"sympathy," linking people like organs in a body. Those who operated telegraph keys forty-eight hours a week, however, had no such experience. In
their fiction, published in journals and anthologies read by other telegraphers, they expressed their frustration and disillusionment with instantaneous electronic communication. For many operators, telegraphy became a
Tantalus torture, always promising intimacy but offering no real human
contact. It provided only very limited knowledge, and its pleasures often
arose not from the information it delivered but from the facts it could not
transmit. While telegraphic engineers were inspired by the nervous system,
operators' responses to telegraphy were much closer to physiologists' warnings about nerves. Those for whom telegraph keys became electronic prostheses knew just how unreliable these new "nerves" were.

Lightning Flashes and Electric Dashes, a collection of stories and poems
by telegraph operators, offered readers exciting dramas built around tenuous telegraphic connections. Compiled in 1877, its tales describe relationships entirely dependent on electronic communication. They would have
been read mainly by other operators in search of moral support. Socially
isolated, expressing themselves in an argot unique to their own set, the
fictional telecommunications workers depicted here bear a striking resemblance to twentieth-century hackers (Standage 64–65). The stories of *Lightning Flashes* expose the conflicting feelings of alienation, power, eroticism,
and paranoia bubbling in those who spend their lives communicating over
the wires. Like Helmholtz's brain, these operators can never be sure what is
happening at the other end of the wire.

J. M. Maclachlan's "A Perilous Christmas Courtship; or, Dangerous
Telegraphy" dramatizes this uncertainty in a simple, direct way, playing the
hacker's power to communicate against the inability to act at a distance. As
the story opens, the Scottish narrator depicts himself as an anonymous
figure in the London throngs. Sensing his isolation, he reflects: "There was
I, a telegraphic idler whose train had rushed him too soon officeward,
among that ever changing crowd" (65). Physically surrounded by strangers,
he feels much less in touch with other people in the London crowd than he
does exchanging thoughts with distant bodies over the wire.

With half an hour to kill, the narrator encounters his old coworker
Frank, "the quondam crack telegraph operator . . . my Colossus of Wires!"
(65). Over a drink, Frank begins explaining how he came to leave telegraphy and, consequently, to prosper. Proud of his newly acquired wealth and
social position, the narrator's friend refers contemptuously to the eighteen
shillings a week he once earned for his "dot and dashing labors" (65).
Although he has abandoned his career as a telegrapher, Frank has incorpo-

rated the operator's mentality to the point that it is reflected in his narration. Interrupting his lengthy description, he tells his friend, "But let us abbreviate, as we used to say on the wire" (66). Face to face with his former coworker, he still associates intimate communication with telegraphy.

Frank opens his rags-to-riches story with an erotic image that runs through all the *Lightning Flashes* tales: that of the beautiful, highly skilled, and highly sexualized female operator who loves to play with words over the wire.[10] The first time he visited the London office, Frank recalls:

> I was led, nervous and bashful, through rows of tittering and whispering young lady telegraphers, who seemed to have no object on earth second to that of rendering a provincial youth awkward and shy. Arriving at "Gb" wire, my *gaucherie* forsook me as if by magic. The lady seated at the key burst upon my dazed senses like a vision of transcendent glory and heavenly beauty. (66)

The rows of beautiful women, arranged like cells in the cortex, simultaneously excite and intimidate him, their function as operators heightening their attractiveness as Frank thinks about everything that their agile fingers can do.

Hired for their lower pay requirements as well as their purportedly superior manual dexterity, female telegraphers were common in the 1870s (Standage 134). Managers of telecommunications networks preferred hiring women because they cost so much less than male employees. When female workers married, they were forced to retire, so that they never rose very high in the pay scale and never drew pensions (Moody 56; Thomas). Frequently, too, female candidates came from a higher class background than young men and were better readers and writers (W. Stone 244). Women, furthermore, were thought to have "a good system of involuntary muscles" and to be "faster than men in light rhythmical activities" (Stubbs 266). This belief in women's superior manual skills and lack of agency in their movements encouraged employers to see them as natural workers, even as machines, while regarding men as women's natural managers.[11] By 1897, women constituted 33 percent of London's telegraphers and 55 percent of its counter clerks (Moody 56).

In 1877, part of the excitement of receiving signals from an unknown hand came from not knowing whether its owner was male or female, a fact that led to many flirtatious exchanges. Even Frank, who knows the gender of his new acquaintance in London, is excited at the prospect of communicating with her, for he knows nothing about her beyond her physical appearance. Enraptured by Violet, his communications goddess, he begins cabling her from his office back in Scotland. At first Violet arouses him with her inaccessibility, for "unlike most other telegraphic ladies, she d[oes] not seem to care to 'do a flirt' on the wire" (66). The freedom of anonymity inspired many telegraphers to play erotic games when the lines were not otherwise occupied, and part of the cultural stereotype of the sexualized operator was her reputed love of suggestive banter and telegraphic mischief. Violet proves properly, frustratingly reserved, but with persistence, Frank "so far improve[s] [the] slight 'wire' acquaintance as to get an introduction to her family," and the two become lovers (66).

Because of the four hundred miles separating them, Frank soon finds that his relationship must be restricted to a series of electronic dispatches. His reflections on their dot-and-dash lovemaking convey another feeling expressed throughout the *Lightning Flashes* collection: a deep frustration with the limits of electronic communication, which only stimulates the desire for physical presence by offering fleeting, tantalizing contact:

> When I returned, sedate, but not unhappy, to the Glasburg end of Violet's wire, such a burning stream of affection, solicitude, and sentiment flowed over that senseless iron thread . . . that I often thought, when our words grew warmer than usual, that the wire might positively *melt,* and so cut the only link that bound us in love together! That link was over four hundred miles long, . . . and yet we seemed as near to each other as if 'twere only a clothes line! (66)

Indeed, by the time *Lightning Flashes* appeared, one did not need to be an operator to appreciate the unreliability of telegraphy. "Freaks of the Telegraph" (1881), which appeared in the popular British journal *Blackwood's Magazine,* reminded readers that "the telegraph is not always, or to everybody, the unmitigated boon and blessing enthusiastic admirers have represented it to be. . . . There is always more or less uncertainty attaching to a telegram, both in regard to the length of time it may be on its journey, and in regard to the way in which the wording may be reproduced" (468). The more Frank communicates with Violet, the more aware he becomes of all that he cannot communicate. Using this uncertainty to create narrative tension, Frank's story explores just how much—and how little—a telegraphic wire can influence actions four hundred miles away.

To distinguish their parasitic communications from the official dispatches for which the line was built, Frank and Violet develop a code within a code. Depending entirely upon private signals, their on-line relationship progresses fitfully as they seek expressions of love their system cannot convey. On Christmas day, a very glum Frank is on duty in his Scotland office when "the London instrument, which had laid inert and silent nearly all that quiet day, to my intense wonderment and surprised delight suddenly chirruped forth 'F.'" Recognizing their private signal, he asks himself, "Could it be my darling at the other end of the wire?" (66–67). Sure enough, Violet has bribed a porter to enter her company's colossal, empty London office and send her lover a private Christmas message. The system is down, but their private, forbidden communications continue.

While dots and dashes can never transmit the touches Frank craves so desperately, they convey a great deal about his partner's emotions through the nuances of her personal touch on the key. Technological communications, like those sent directly by the body, have a distinct personal rhythm, and from the duration and spacing of Morse's dots and dashes, an experienced operator could detect not just the sender's sex and identity but her personality and mood (Standage 130). As is the case physiologically, the interpreting brain could make sense of incoming signs by reading them relative to one another.

While Violet is in the abandoned office, robbers break into the building, and Frank suddenly "hears" a change in her style. He perceives

> a tremulous line, ending in a series of unconnected dashes and dots that seemed to shiver on the armature, just as if the hand that held the key wavered and shook with some strong emotion, and endeavored vainly to form characters almost mechanically. (67)

Here Maclachlan makes it clear that the nerve between London and Scotland transmits not just information but emotion as well. Keeping her head, the beleaguered Violet cables Frank for help. Restricted by problems of "real space"—and real economics—he runs across the street to a competing telegraph company and sends a frantic message to their London office, which, unlike his own company's, is open on Christmas and is located just two blocks from Violet's.

But once he has wired for help, Frank realizes his actual powerlessness. Well aware of his separation from Violet, he can only wait as he envisions his beloved violated or murdered by the thieves. In despair, Frank curses the tantalizing instrument that never offers quite enough contact: "forgetful of the four hundred miles which separated me from the cruel wretches who threatened all I held dear, I desperately shook the fatal instrument in a paroxysm of impotent rage" (68). While the telegraph promises intimate contact, it tortures its operator by failing to provide the physical presence it suggests.

Thanks to the communications power the wires do provide, help arrives in time, but Frank is unimpressed with these extensions of his nerves and resolves to free himself from this Tantalus torture forever. Interestingly, the story upholds his rejection of telegraphy by rewarding him with a wife (Violet), a child, and a lucrative new career, possibly because Maclachlan shares his frustrations. Through Violet's family, Frank obtains a position on the stock exchange, where he subsequently "prosper[s] amazingly" (68). But by 1877, of course, the British stock exchange owed its burgeoning life to the same tantalizing telegraphy abandoned by Frank (Standage 170). Although this one Scottish telegrapher escapes the tenuous relations offered by the net, his entire society is coming to depend upon these relations for its socio-economic survival.

Like "A Perilous Christmas Courtship," Charles Barnard's tale, "Kate: An Electro-Mechanical Romance," depicts people's need for intimate contact and private communication. In particular, it suggests the pleasures and dangers of communicating privately in a public web. In this story, the problems of personal contact in the telegraphic network merge with the closely related ones offered by the railroads, for the ingenious and frustrated lovers are a telegrapher and an engineer.

Barnard introduces Kate and John, the main characters, through their intimate relationships with machines, and he conveys the mechanical aspects of their courtship through highly eroticized descriptions of their technological counterparts. The opening line of the story, "She was a beauty," delivers an ironic jolt when the reader realizes that this "beauty" is not a woman, but John's locomotive:

> From head-light to buffer-casting, from spark-arrester to airbrake coupling, she shone resplendent. A thing of grace and power, she seemed instinct with life as she paused upon her breathless flight. Even while resting quietly upon the track, she trembled with the pulsations of her mighty heart. . . . She seemed long and slender like a greyhound, and her glistening sides, delicate forefeet, and uplifted head were suggestive of speed and power. (53)

Before John ever meets Kate, he is married to his engine, which the author presents both as an eroticized mate and as an extension of his own body. Once he discovers the young telegrapher, a reference to the engine's "forefoot . . . daintily thrust out in front" aligns the locomotive with his prospective human mate, who takes a brief ride with "one little foot steadied against the boiler" (53, 55). When John shows Kate his mighty engine, the narrator tells us, "the little electrician [i]s charmed," thinking, "what a magnificent machine he ha[s] beneath him!" (55). John loves his engine, and the author conveys his excitement over Kate by expressing each passion in terms of the other.

Like the operators in "A Perilous Christmas Courtship," Kate feels socially isolated and finds intimacy only through electronic exchanges. In her rural station office,

> There was a sunny window that looked far up the line, and a little opening where she received the messages. She viewed life through this scant outlook, and thought it very queer. . . . Sometimes between the trains the station was quite deserted, and were it not for the ticking of the clock and the incessant rattle of the fretful machine on her desk, it would be as still as a church on Monday. At first she amused herself by listening to the strange language of the wires, and she even made the acquaintance of the other operators. (54)

When Kate listens to the communications of others, their "tongues of brass" strike her as mechanical and uninteresting until she hears a sophisticated female "voice" that charms her ear. Like Frank, Kate has a sensitive ear for the language of the wires, and she recognizes one woman (Mary) in a distant city as a "lady." Gradually, as Kate's and Mary's personal messages slip through between the official ones, the circuit connecting the telegraphers lets them become intimate friends.

Sadly, in their attempts to communicate face to face, Kate and John are nowhere near as successful as the two women over the wire. As luck would have it, the lovers are continually thwarted by dispatches that reach Kate's station just as John arrives, chaining her to her key. Unable to exchange thoughts in person, they are left with only the vaguest, most unsatisfying signals. "Ah!" exclaims the narrator, "a dress fluttering in the door-way. . . . With both hands on the throttle-valve, the engineer leans out the window. A handkerchief is quickly flirted in the air" (54). Restrained not just by physical and economic circumstances but by their fear of discovery, Kate and John find themselves unable to communicate. Despite their obvious attraction, the engineer and telegrapher do not want their mating signals to be read by others, and they long for a system in which they can "talk" more privately. Simply to communicate is not enough. Like the other characters in the telegraphers' stories, they long to exchange thoughts through signals known to them alone.

When the frustrated lovers finally do have a private conversation, it is about how to communicate. At first, Kate suggests that John whistle to signal her when he passes. Recreating Helmholtz's argument about the possibility of human knowledge, she persuades him that a single sounding of the whistle has no meaning in itself but that a series of blasts becomes informative because of the time intervals between them. When he gives her only a single sounding, she explains, "I cannot tell your whistle from any other, and so, I sometimes miss seeing you." Seeing her point but unwilling to make his communication public, John replies, "if I gave two whistles or three, they would think it meant some signal, and it would make trouble" (55). The urge for intimate contact proves too great, however, and when Kate teaches John how to blast out her name in Morse code, he learns the signal eagerly. For a time, the lovers are content as John blares out this mechanized mating call twice a day.

Kate still fears discovery, however, for their code is shared by an entire community of hackers (Standage 129–31). Late one evening, she signals Mary and confesses her worries to her friend. Like Maclachlan, Barnard makes it clear that even when one communicates through a series of electronic breaks, it is possible to have a personal style:

> At once the two girl friends were in close conversation with one hundred miles of land and water between them. The conversation was by sound in a series of long and short notes—nervous and staccato for the bright one in the little station; smooth, legato and placid for the city girl. (56)

By comparing the flow of signals to music, the author shows quite aptly how patterns of sound can suggest moods and personality traits. To help her country friend, the clever Mary suggests a more daring but more secure way of communicating: Kate and John must rig their own private, parasitic circuit. At first Kate hesitates, for such tampering violates company rules, but eventually she decides to build the circuit when two itinerant operators hear John's "Titanic love-signal" and ask each other "Kate! Who's Kate?" (57). There can be no private messages when everywhere there are trained receiving ears.

As the determined lovers set out to build their private circuit, Barnard describes the construction in such detail that the story can only be read as a how-to manual for hackers. Using an abandoned wire, a pickle-jar battery, and eleven dollars worth of electronic gear, Kate and John build a circuit that will ring a bell in her office whenever his engine passes a pair of trees several miles away. Because of a special rod he has attached to his engine, the bell sounds for his train alone. Thanks to modern communications technology, Kate and John are on the brink of achieving the intimate contact for which they have so sorely longed.

But as Kate waits anxiously to see whether their new apparatus will succeed, her own overloaded neural circuits create chaos in the technological ones. Most telegraphic errors, observes another writer in 1881, can be traced to the living portion of the apparatus:

> it is the 'personal equation' which has to be allowed for. The human element plays so considerable a part in matters telegraphic, that the human propen-

sity to err finds proportionately wide scope. . . . It evidently to a great extent depends on the turn of mind of the operator which way [the messages] come out. ("Freaks of the Telegraph" 468, 475)

Describing Kate's distraction, Barnard suggests the interdependence of physiological and mechanized circuits, which have their interface in the beleaguered heroine:

> It was very singular how absent-minded and inattentive the operator was that day. She sent that order for flowers to the butcher, and Mrs. Robinson's message about the baby's croup went to old Mr. Stimmins, the bachelor lodger at the gambrel-roofed house. (59)

Like the apparatus to which she is attached, Kate is a communications device, of sorts, and both the technological and the physiological systems are susceptible to breakdown and overload. When either system is overtaxed, messages can be misdirected, and communications can go awry.

As Kate's poor performance makes clear, the human portions of telegraphic circuitry were often its weakest links. Because the railways depended upon telegraphy for their signals, they were open to error in any way that their human machinery was vulnerable. As Barnard's narrator puts it, "life [is] an iron road with dangers everywhere" (54). One evening a group of hungry railroad company directors stop at the station for dinner, leaving their train on the main track. Unexpectedly, Kate's private signal sounds, warning her that John's engine is approaching, and she cries to the directors to move their train to a siding. Thanks to her warning, the deadly collision never occurs, but the grateful directors ask how she knew John's train was coming. When she confesses, they call a meeting on the spot and declare that:

> whereas, John Mills, engineer of engine Number 59, of this railway line, erected a private telegraph; and, whereas he, with the assistance of the telegraph operator of this station (I leave a blank for her name), used the said line without the consent of this Company, and for other than railway business: it is resolved that he be suspended permanently from his position as engineer, and that the said operator be requested to resign. (61–62)

If the company is to control the communications system on which it depends, it can permit no private use of its circuits, no unauthorized local power centers in its information net. While the directors remove Kate and John from their web, they reward John for his initiative and technical skill, appointing him chief engineer at a new repair shop. There is no mention of a job for Kate.

Like "A Perilous Christmas Courtship," Kate and John's "electro-mechanical romance" ends with the substitution of presence for absence, of physical for electronic contact as the operator and engineer bond with each other instead of their mechanical devices. With desire fulfilled, there is no further need for telecommunications, technical ingenuity, or narration itself. "Handiness," the tale implies, is best devoted to the public good, not to the construction of private circuits. Private communications conducted over public networks will always be discovered in the end. In its closure, the American tale echoes the Scottish one in privileging physical presence over long-distance communication. This is not the case for all the stories in the collection.

A third and final tale, L. A. Churchill's "Playing with Fire," represents the pleasures and risks of nineteenth-century telecommunications with much deeper irony. Like the stories of Frank and Kate, it explores the frustrating limitations of telegraphy, but it far exceeds them in suggesting how these restrictions can be used to creative advantage.[12] In this story, the heroine, Rena, is a skilled, eroticized dispatcher who in the opening paragraph "thr[ows] back her curls in an energetic manner with her left hand while she open[s] her key with the right" (69). Rena, however, lacks Kate's and Violet's mating instincts. She may be "the best looking operator on the line," but for her, the wires come first (69). Rena draws her identity from her profession as a telegrapher, and she loves to play games with her key.

In "Playing with Fire," the problem of knowing to whom one is writing merges with broader cultural anxieties about people's abilities to know one another in general. As suggested in the other stories, physical presence does not necessarily facilitate communications. The inability to know the person one is addressing, these authors imply, is not a problem unique to telegraphy. By complaining about the shortcomings of telegraphic communication, they articulate the difficulties of communication on any level. Despite widely held assumptions about class and gender, one can never know the real thoughts of one's neighbors. In this operator's tale, it is Americans' assumptions about male and female behavior that come under attack.[13] Woven into what might have been a traditional story of love and marriage are subversive avowals of on-line flirtation for its own sake.

"A merry body, and fond of fun," Rena resolves to trick a female dispatcher in a nearby office by wooing her over the wire as a man. "I must do something to keep from stagnating in this dull office," she vows. "Yes, I will sign a man's name and fool her in grand shape" (69). Like Kate and Frank, Rena is a seasoned operator, and she communicates with the puns and abbreviations of nineteenth-century hackers (Standage 130). She tells the other telegrapher, whom she knows only as "Bn," that her name is Isaac:

"My name is Isaac, but I sign 'Ic.'"

"U are joking, I fear. Is your name really Isaac?"

"I am not, truly. If you ever want me call 'Ic.' . . . I saw u sitting near ur window the other day while I was passing there on the train, and . . . I compared u to a white rose I held in my hand." (69)

While their ears detect only clicks and pauses, the two operators' abbreviations rely upon the sound of the spoken language—itself a system of signs—that their signals supposedly represent. Although telegraphers work with dots and dashes, it is their delight in spoken language's confluences that drives the game. Rena's correspondent seems pleased by this compliment, and Rena thinks, "If I can make her think I am really and truly a man, I will have some fun of the first water. . . . I fancy she liked being told she looked like a rose" (69). Happy to be desired by this unknown woman, Rena continues her suit, and the female telegrapher responds eagerly.

One day, Rena receives a letter requesting an encounter in "real space." The railway timetables have permitted her friend a layover of one hour and twenty minutes at Rena's station, and the on-line lovers can finally meet in the flesh. Regretting the end of her intimacy with the unknown woman, Rena thinks, "I am so deeply interested in her that I more than half wish this Isaac business was a reality" (70). When the train arrives, however, no female passenger disembarks. Instead a "tall, fine-looking fellow" approaches her, asking for Isaac (70). Realizing that the game is up, Rena confesses that she is Isaac but tells him that she is waiting to meet a friend. As an experienced operator, the attractive man asks, "What does your friend sign?" (70). To one who spends her life reading electronic signals, one's personal "sign" is a mark of identity as telling as her own face, and Rena responds, "her office call is 'Bn,' but I believe she signs 'D'" (71). The fine-looking young man, of course, is "Bn." The trickster has been tricked by the same strategy with which she hoped to fool the female operator: she has made false assumptions about the origin of the signs she is reading.

Like a Shakespearean comedy, this operator's tale offers a socially sanctioned narrative closure after experimenting with traditional gender roles. "Bn"—who believed that he was intimate with another male operator and thoroughly enjoyed it—confesses: "often have I wished I *was* Miss Dwinnell, if Isaac would care for me as he seemed to care for her. But things are now just as they should be, and if you do not consider the action too abrupt, I would like to ask you to be my wife" (71). Rena, however, refuses his offer, and Churchill rejects the restorative ending that the reader expects. "Home love is not for me," declares Rena. "I have duties you know nothing of—a purpose in life which I must work out alone" (71). More devoted to her key than to family values, she dedicates her life to telegraphy. In an epilogue, Churchill kills off the stubborn heroine, perhaps in consternation for her denial of traditional gender roles. "A few years later Rena died," we read. "Her work was done" (71). Apparently, even Rena's invocation of the Calvinist work ethic cannot overcome her rejection of domesticity. Churchill's harsh conclusion, however, is as unconvincing as that of Choderlos de LaClos's *Les Liaisons Dangereuses,* the classic tale of deceitful communications gone awry. In the telegrapher's story, as in the French novel, the real life of the tale is in the perverse courtship, not in the severe dénouement. Churchill's story suggests that if one uses the new communications system to create an "optical illusion," establishing a new, fictional identity, the physical world will eventually destroy the false image, but what fun one will have while it lasts!

For nineteenth-century operators, not knowing who was at the other end of the wire could be disturbing, even frightening, but it was also liberating and exhilarating. Along with the anxiety of not knowing came the pleasure of not being known, so that telegraphy's tenuous relation with the real world simultaneously empowered and disempowered the lonely operator.

Chapter 5
Two Telegraphers Unhappy with Their Nerves

Intimately connected to the keys and wires that provided their raison d'être, nineteenth-century telegraphers viewed their electronic tongues and nerves as extensions of their own bodies.[1] The contact provided by these prosthetic nerves and sensory organs, however, proved as tenuous as that of late twentieth-century E-mail relationships (Standage 209). In stories about telegraphy, one reads not just the concerns of communications engineers but the concerns of scientists investigating the nervous system. On the one hand, operators found it exhilarating to receive coded messages from distant cities, messages that only they could decode and understand. Like Helmholtz, however, telegraphers suspected that their "nerves" were conveying pitifully little information about the real world. While it was thrilling to be in touch with unknown hands at distant keys, one could never truly know the numerous people with whom the new nerve network connected one. The organic and technological communications systems performed the same functions, and they shared the same weaknesses.

The Struggle to Communicate: Ella Cheever Thayer's *Wired Love*

Unlike the stories of *Lightning Flashes,* Ella Cheever Thayer's novel *Wired Love* (1880) draws the lay reader into the eroticized telegraphic circuit.[2] Thayer, an advocate of women's rights, depicts a telegrapher not so much to explore a new technology as to convey working women's struggle for self-expression and economic survival. Soon after the publication of *Wired Love,* Thayer would write *Lords of Creation: A Suffrage Drama in Three Acts* (1883), a play challenging negative stereotypes of suffragettes (Davidson and Wagner-Martin 862). Written by and for operators, the stories of *Lightning Flashes* formed part of the telegraphers' support network, providing an opportunity to share frustrations. *Wired Love,* on the other hand, is intended for all working women; the calling of telegraphy is merely representative.

While directed toward a broader audience, Thayer's novel is clearly about communication. Like the protagonists of the *Lightning Flashes* stories, her characters challenge public optimism about bonds forged by the new technology, and their assessments of its epistemological worth are closer to those of contemporary physiologists. Offering unfulfilled promises of knowledge and intimacy, the telegraph described in these works leaves its users open to fraud when they have too much faith in its possibilities. By associating the telegraph with the plight of working women, Thayer makes readers evaluate frustrated communications in their own lives. In her popular novel, the telegraph presents an exciting new means of self-expression, but like other technologies that offered women employment, its promised liberation proves all too limited.

In some respects, Thayer's novel seems to have been written for telegraphers. The dedication and concluding lines are in Morse. Generally, however, Thayer takes pains to describe equipment and define terms for readers unaccustomed to sending their own messages. In the opening scene, protagonist Nattie Rogers struggles to transcribe a rapidly incoming message. Repeatedly, she is interrupted by an old woman who asks her, "Do you take [the messages] entirely by sound? . . . Do you have a different sound for every word, or syllable, or what?" Here, Nattie's irritated but informative answers keep the reader apprised of what she is doing (12–13). Later, the narrator takes over this informative role, breaking the story's circuit to open up telegraphers' private language to the reader. In the first scene, we later learn, Nattie is being "salted" (intentionally sent a message more quickly than she can take it) by "C," an operator down line who wants to test her telegraphic prowess. Like Churchill's "Playing with Fire," Thayer's novel speaks more convincingly of the delights than the dangers of forbidden communications. It is a how-to book not for hackers but for any readers who feel isolated for social and economic reasons.

Thayer's description of the Morse system affirms Siemens's early impression of it as a religion or cult. The narrator explains that there is a "sort of freemasonry between those of the craft"(95). As one who hears voices that others cannot, the telegrapher looks like a lunatic to the uninitiated, laughing—apparently—at nothing, suddenly turning violently active, and then returning as quickly to rest (Thayer 78). When Quimby, the novel's buffoon, hears that Nattie has "made a new [acquaintance] on the wire," he thinks first of a tightrope and then of a telephone, which had been available to the public for three years at the time (Thayer 38–39; McCallum). "We don't need telephones," Nattie replies scornfully. "We can talk without.

. . . And . . . no one but those who understand our language can know what we say!" (39). Quimby, who fails to learn Morse, likes the telephone because anyone can use it, but Nattie prefers telegraphy for the opposite reason. For her, Morse is a secret language that can end isolation and open new, forbidden connections, but only if one is willing to learn it.

In Thayer's obvious attempts to win sympathy for Nattie, one hears the same coextensive, contradictory discourses that clash in "Playing with Fire." On the one hand, Thayer's novel is another version of "the old, old story," as Thayer—or her editor—indicates on the title page.[3] It is a love story in which an active male suitor woos a reluctant female, wins her favor, loses it, and then finally regains it, a tale that will almost certainly end with a happy marriage. Nattie, a young, single telegrapher in a small Western city, "meets" a sympathetic operator over the wires, a man she knows only as "C." As they exchange intimate thoughts, they quickly grow fond of each other, until an unscrupulous operator who has overheard their messages visits Nattie in the flesh. Falsely claiming to be "C," he tries to beat her on-line lover to the physical consummation. Repulsed, Nattie breaks off all communication with "C," who then moves to her city to woo her face to face. Using his real name, Clem Stanwood, he begins his suit from scratch, slowly befriending shy Nattie, who learns his identity only in the end. In the novel's concluding dots and dashes, he fulfills readers' romantic expectations by calling Nattie "my little darling, my wife" (256).

On the other hand, the author does everything possible to show that Nattie is "not the kind of girl to sit down and wait for someone to come along and marry her" (Thayer 28). The protagonist is an active, responsible woman, and her clearly stated ambitions and dreams of getting ahead give voice to a capitalistic discourse more commonly associated with men. Likely to appeal to any isolated, overworked reader with dreams of intimacy and fame, Thayer's novel is truly the " 'old, old story'—in a new, new way."

In its portrayal of female desires, *Wired Love* was most likely to appeal to working women of modest means. Nattie is "not one of those impossible, angelic young ladies of whom we read," we are told, "but one of the ordinary human beings we meet every day" (15). Openly protesting against traditional images of women as helpless, delicate beings, Thayer's narrator remarks that:

> storms of nature, like storms of life, are hardest to a woman, trammeled as she is in the one by long skirts, that will drag in the mud, and clothes that every gust of wind catches, and in the other by prejudices and impediments of every kind, that the world, in consideration, doubtless, for her so-called 'weakness,' throws in her way. (175–76)

Thayer's portrayal of Nattie involves no weakness, only a very reasonable, well-justified frustration. In what little we hear of Nattie's personal appearance, it is suggested that she is handsome rather than pretty. "Pretty people seldom know very much," comments the narrator, "but to be handsome, a person must have brains; an inner as well as an outer beauty" (43). Nattie's intelligence flashes out in her cravings for fame as a writer and her frustrations with her dead-end job.[4] At eighteen, she is working as a telegrapher because she must support herself financially. Her father has died after failing in business, and her mother can barely support her two younger children. As a middle-class woman leading a working-class life, Nattie stands to win readers from both classes.

Although she is a woman, Nattie frequently expresses the "rugged individualism" associated with young men like Andrew Carnegie or Horatio Alger. Upset that she is "making no progress towards her only dreamed of ambition," Nattie confesses to her friend Cynthia (Cyn) Archer, "I cannot be content with a mere working on from day to day, in the same old routine, and nothing more. . . . I have attacks of energy" (68, 119–20). Cyn, who with her small independent income shares Nattie's poverty and middle-class roots, is "determined to fight [her] way up" as a singer and actress (68). She and Clem Stanwood ("C"), Nattie's on-line lover, express the same ethic of self-reliance and individual responsibility. When Nattie tells Clem, in a weak moment, that "it seems so hopeless. There is no opening anywhere," he replies, "if the world is a closed oyster, we must open it" (190). Cyn, for her part, scorns the idea of fate, asking, "what incentive would we have to any effort, if we were sure everything was marked out for us in advance?" (221). Thayer depicts a world in which women can dream of success and self-sufficiency, but where they encounter every possible barrier.

Thayer's novel is openly hostile, on the other hand, to any notion of inherited privilege. This attitude emerges in the descriptions of its pathetic villainess, the "nosy" landlady Miss Kling. Feeling infinitely superior to her boarders, the spinster "boast[s] of a certain 'blue blood' inherited from dead and gone ancestors" (26). Miss Kling's cat, The Duchess, periodically appears in the characters' rooms, as does Miss Kling herself, in her unending efforts to see that her lodgers meet her moral standards. When depicting the intolerant landlady and her boarders, Thayer juxtaposes the young people's desire for mobility with the fallen aristocrat's longing for a rigid, traditional social order. Throughout the novel, their attempts to communicate challenge Miss Kling's efforts to restrict contact between those who resist that order.

At the same time, the main characters and their sympathetic narrator express a very marked scorn for the lower classes. As voiced by Quimby, who occasionally expresses himself aptly, Nattie's greatest anxiety about her on-line lover is not that he may be dangerous but that he may be a "soiled invisible," a man with dirty cuffs (55). When a brazen, red-haired man comes to her office claiming that he is "C," the romance founders, temporarily, as Nattie reacts in horror to his "vulgar assurance" (97). "C" is "interesting, witty, and gentlemanly," she thinks, whereas this creature is "a musk-scented being of greasy red hair, cheap jewelry, and vulgar manners" (97). Just as Thayer's novel both challenges and incorporates traditional notions of romance, it endorses some of the traditional class prejudices it aims to overthrow.

Whatever the characters' ideologies, the grim reality of their lives places them solidly in the working class. Despite the apparent glamour of telegraphy, Thayer reveals, it is really a monotonous, poorly paid, dead-end job. Yet while telegraphy epitomizes the dehumanizing working conditions of so much nineteenth-century employment, it allows frustrated workers to communicate with other people in the same predicament. Nattie describes her office to "C" as "a long, dark little room, into which the sun never shines, . . . [with] a high stool, desk, instruments—that is all—Oh! and me" (20–21). Like any factory worker, she merges with her machinery.

When Cyn first sees Nattie's office, she remarks on the "confinement" of the place, but Nattie sees openness, telling her, "there always is someone to talk with 'on the wire'" (52). In a moving passage, the narrator conveys how much the wires mean to Nattie, who "live[s] . . . in two worlds" (25). From her room in Miss Kling's boardinghouse, she can see only a shed, an ash barrel, and occasionally some flapping sheets. From her office, however, "she could wander away, through the medium of that slender telegraph wire, on a sort of electric wings, to distant cities and towns. . . . Although alone all day, she did not lack social intercourse" (25). Quick to understand Nattie's situation, Cyn teases her with a drawing of a man and a woman, each alone in a tiny room, joined only by a wire. Overhead, Cupid is hovering. Later, Cyn jokes that if Nattie were alone on a desert island, one could just "set up a telegraph wire, and then she would need nothing more" (222). While Nattie leads a life of drudgery and loneliness, her wires fulfill a hunger common to many nineteenth-century working women. They offer a life "rich . . . in acquaintances" (Thayer 38).

In many ways, *Wired Love* is a novel about hunger, and in one touching scene the young women's desire bursts out as a real, physiological craving for food. Telling Cyn of her ambition, her longings for fame as a writer, Nattie concludes in despair that she can't even afford a decent meal. To comfort her, Cyn suggests that they pool resources and plan a secret repast. In great detail, Thayer describes their efforts to find money, supplies, and finally enough cracked dishes and wobbly furniture to serve their subversive feast. More a pitiful reflection than a parody, their dinner party suggests the role they would be playing if they had assumed their expected positions as young bourgeois wives. But even in this literal, physiological expression of their desire, the working women are denied the satisfaction of their appetites. Just as they are about to tuck into a dinner big enough for ten, Quimby bursts in with an unknown young man—Clem, who is the distant lover, "C." Clumsy as always, Quimby sits down on their precious pies, which have been placed on a chair due to lack of table space. Norton, an artist, joins the women as well, protesting "the idea of you two girls proposing to selfishly enjoy such a feast all alone," and the three men devour a good deal of the food (142). Clem declares that he has not had so much fun since he used to steal eggs as a boy. Like the lovers' illicit messages, the meal is a stolen pleasure.

Reflecting several of the other characters, Nattie must satisfy her hunger through secret communications not just because of surveillance but because face-to-face encounters present an even greater challenge. In Quimby's perpetual failure to express himself, the klutzy clerk serves as Nattie's alter ego (McCallum). Desperately in love with Nattie, Quimby holds a job similar to hers, "ek[ing] out his meager income by writing in a lawyer's office" (85). Like his beloved, he is "always making mistakes" (81). Thayer uses the same word, "bashful," to describe both characters. Once Nattie has learned that "C" is Clem Stanwood, the "tall, fine-looking young man" that Quimby unwittingly invited to dinner, she becomes "strangely timid" and can barely speak to him at all (130, 157). Eager to continue his courtship, Clem begins tapping out messages to Nattie, first with a pencil, then later—in her presence—with the key in her office, for she confesses that she can "talk more easily" over the wire (172). It is not just for himself that Quimby speaks when he cries, "Oh! I can't express myself! It all comes upon me with a rush when I am alone, but now, at this supreme moment, I cannot tell you how I a—" (89). Quimby's first attempt to propose to Nattie collapses utterly, and in his second, he is accepted by homely Celeste Fishblate, who has replaced his beloved in a dark room.

In Miss Kling's lonely isolation and quest for a soul mate, even the mean-spirited landlady bears a certain affinity to Nattie. "Why should others be happy," she wonders, "when she had not found her other self?" (103). The old maid's determination to quash all communications among others results from her own inability to find a satisfying connection, and she is left "sitting lonely by her fireside, and pining for her other self" (169). Thayer's novel tells about high-tech communications, but it is really about people's inability to communicate.

Even if the characters could overcome the inherent difficulty of expressing their thoughts, Miss Kling's perpetual surveillance and the socioeconomic system under which they live would make communicating extraordinarily hard. Introduced initially simply as "the nose," nosy Miss Kling embodies a much older, more traditional social network for transmitting intelligence: the web of gossip. With her enormous beak, Nattie's landlady can smell anything amiss and is "addicted to sudden and silent appearances, much after the manner of materialized spirits" (27). There is nothing spectral about Miss Kling, however. Like Nattie and Clem, she spends her life gathering and transmitting information, but whereas they use their keys and wires, she uses her clinging flesh. Even simple-minded Celeste Fishblate remarks that the landlady "does *Kling* in a way that is not pleasant" (217). A more primitive, animal-like sensory organ than the eye or the ear, Miss Kling's nose suggests the body's own systems of detection. When her nose is poked into Nattie's office in the opening scene and its owner's penetrating questions interrupt the flow of information between Nattie and "C," Thayer juxtaposes the intrusive organ with the busy woman's telegraph key.[5]

Unable to stand the possibility that anyone could be better informed than she, Miss Kling tries to freeze the flow of information. She hopes to create a perfect Panopticon in which she is the only privileged viewer—or smeller, as the case may be. The old landlady objects violently to Cyn's public appearances as a singer, declaring that "public characters are not to be trusted" (64). If she could have her way, neither bodies nor thoughts would circulate freely. With her hypersensitive nose and clinging ways, she suggests the restrictions that corporeality imposes on the free flow of communication.

In *Wired Love,* the modern technology of the telegraph presents a delicious means of subverting such physical and social obstacles to communication. As a new, extended ear, tongue, and nervous system, the telegraph provides an exhilarating way to exchange thoughts without detection, a vast improvement on the organs furnished by the human body. Originally, Clem tells Nattie, his father tried to force him to become a doctor, but he felt "a repugnance to knowing anything more than absolutely necessary about 'the ills that flesh is heir to'" (160). Clem is disgusted, perhaps frightened, by the body's vulnerability, but he embraces the new technology that mimics its functions. In his college dormitory—like Miss Kling's boardinghouse, a Panopticon for controlling the young—he learned telegraphy so that he could "talk" freely with the boys who had strung wires from room to room.

The spirit of telegraphy, he quickly learns, is the free flow of information. In the argot of telegraphers, an incompetent operator is a "plug" (Thayer 98).

Once Nattie, Cyn, Quimby, and Clem are settled into the same carefully watched house, Clem wires their rooms so that they can talk freely throughout the night. Unbeknownst to Miss Kling, they communicate right "under her very nose," and Cyn, who eventually learns Morse, is thrilled that they can "get the better of that argus-eyed Dragon" (167, 176). From the outset, Nattie has seen the potential of her code, telling Quimby excitedly, "we talk in a language of dots and dashes, that even Miss Kling might listen to in vain" (39). Barely able to communicate with Clem in the presence of vivacious Cyn, Nattie loves their late-night exchanges because "*nobody* [can] share" them (175). The telegraph appears to offer unlimited opportunities for private communications in a rigid, tightly controlled world.

In reality, though, the telegraph provided only very limited possibilities for exchanging confidences. Telegraph companies forbade the circulation of personal messages among operators, so that Nattie's and Clem's effusions could have cost both of them their jobs. The structure of nineteenth-century webs, especially in rural areas like the one depicted in *Wired Love*, ensured that no communication was ever really private. Along each line radiating outward from a central office, every operator listened to each dispatch destined for any station along the line. Only in this way could operators know when the wire was free so that they could transmit their own messages (Standage 63–65).

In *Wired Love*, during almost all of the lovers' exchanges, sarcastic or irritated operators in other offices interrupt their on-line lovemaking to offer advice or simply to demand that they "get out!" (45). It is one such operator who conceives the devilish notion of impersonating "C" in the hope of winning some physical affection from Nattie. When the impostor tells her, "I think we astonished some of them on the wire with all the stuff we had over," it occurs to her for the first time how many people may have been listening (99). When Nattie is repulsed by the interloper and the lovers' relations temporarily break down, we learn that "the sudden cessation of the intimacy between 'C' and 'N' was a theme of much surprise and bantering comments along the line" (114). As in Barnard's story of Kate, Morse code at first promises a means for secretive personal communication, but it is a secret to which an entire community is privy. With its own surveillance and webs of power, the telegraph network recreates the tendencies of the power network it promises to subvert.

Thayer reinforces this impression by using telecommunications devices as metaphorical vehicles for human relations in general. "Fate is a sort of switch-board," we are told, "and a slight move will switch two lives onto wires far asunder, even as the moving of a peg or two will alter everything on the board that shows its power so little" (93–94). Later, Cupid is compared to an incompetent telegrapher in a central office, one who has "switched everybody off onto the wrong wire" (194). By the early 1880s, a number of writers for the popular press were speculating about the playful beings who sabotaged telegrams. A writer for *Blackwood's Magazine* commented that "too often some kind of tricksy spirit, some telegraphic Puck, seems to preside over the destinies of the telegram, with malicious perversity altering the sense, and seeming to take a pleasure in thwarting Man, and playing practical jokes upon him" ("Freaks of the Telegraph" 468). While Thayer's references to fate and mischievous gods run all through her text, Clem's and Cyn's urges for personal responsibility carry far greater weight. It is people, not fate or gods, who are establishing contact, failing to communicate, and sabotaging one another's communications. When frustrated Nattie compares herself to a " 'line man,' who goes up and down to find why the wires will not work [but can] not find the 'break' anywhere," Thayer's metaphor speaks mainly of the differences between people's and machines' capacities to communicate (195). A broken wire can be detected and repaired, but a malfunctioning relationship is not so easy to diagnose.

In *Wired Love*, however, it is difficult to distinguish minds and bodies from machines. They perform the same functions, and, like the narrator, most of the characters offer images in which physiological and technological systems merge. In the opening scene, in which "C" calls Nattie for the first time, she expresses scorn for "the possible abilities of the human portion of [the] machinery" in his small country office (10). She envisions the unknown operator as she envisions herself, as a mind that reads and a hand that transmits, two fallible pieces of equipment attached to a telegraph key. When the narrator describes Celeste Fishblate's irascible father as "exploding in short, but expressive sentences," the telegraphic key comes to mind as an analog for his inordinately noisy body (32). Even Miss Kling, a stranger to telegraphy, knows that "machines do not clatter without a human agency somewhere" (185). When Nattie's on-line relations with "C" break down, she realizes that she misses his "sympathy" (114). While made possible by a wire, theirs is more like a physiological bond, like connections created by nerves in the body.

If her telegraphs act like nerves, Thayer also emphasizes the body's mechanical possibilities, its electrical and magnetic forces. "Explosive" old Fishblate is christened "The Torpedo," even though his deadly flashes of temper fail to discourage Miss Kling, who sees him as a candidate for her "other self" (125). In the novel's climactic moments, the landlady detects the young people's private telegraph and excoriates the women for allowing such intimate contact. Cyn, however, shocks Miss Kling with a "Torpedo-wound" of her own, snapping that the old maid must know all about old females running after old men (250). When Nattie and Clem finally embrace physically, the discourse of telegraphy melts into a description of the physiological forces that make it possible: "Neither words or dots and dashes were needed. Love, more potent than electricity, required no interpreter, and that most powerful of all magnets drew them together" (250–51). When Thayer compares bodies' and machines' capacities to communicate, her metaphors are playful, but they point out fundamental interdependencies. Bodies in her novel communicate best when attached to machines, and machines can communicate only when attached to human hands and minds.

Even as *Wired Love* opens, the reader sees Nattie's mind as an essential part of her apparatus. Like her key, it perceives and translates signals, and it encounters many of the same difficulties in the process. Thayer's very first sentence, "Just a noise, that is all," suggests the mind's role in the telegraphic apparatus (9). At the same time, it brings to mind Nietzsche's chain of metaphors linking human knowledge with the world. To the untutored listener and to most of Thayer's readers, the clatter of Nattie's key is merely a noise. To Nattie's experienced ear, however, it is a meaningful message. As a transmitter, her key turns electrical signals from her brain into electronic dots and dashes, while her brain, a "receiver," provides the final link that makes meaning out of electrical impulses. Her discerning mind must also detect meaning in her customers' misspelled scrawlings ("meat me at the train") before translating them into Morse and making them available to mechanical receivers (43). "Telegraphy is a species of dictation," noted one critic in 1881 ("Freaks of the Telegraph" 469).

Like her circuit, in the opening scene, Nattie's concentration is repeatedly "broken" by Miss Kling's questions. Exasperated, she explains that like her machine, she cannot transmit and receive simultaneously. When she is not distracted, Nattie is a highly skilled reader. The central problem with Thayer's story is Nattie's inability to perceive first that the red-haired impostor is not her "C" and second that Clem Stanwood most certainly is, for these facts are immediately apparent to the reader.

If the discrepancy between Nattie's on-line and off-line reading skills is intentional, as it may well be, it points to one of the most fascinating aspects of communications that Thayer depicts. In her novel, there is no apparent connection between telegraphic language and events in the real world. Just as Helmholtz and Nietzsche wondered what relation mental symbols bore to actual events, and just as contemporary users of E-mail revel in the free flow of virtual language, Nattie speculates that "people talk for the sake of talking, and never say what they mean on the wire" (41). To the bashful above all, telegraphic communication frees one from corporeal encumbrances like blushes and stammers. It offers contact without consequences. By bracketing reality, it invites no end of fictions.

Reflecting on her relationship with "C," Nattie realizes that she has "woven a sort of romance about him who was a friend 'so near and yet so far.' . . . used to the license that distance gave, whether wisely or unwisely, [she] had never thought it necessary to check the familiarity" (92–93). Nattie assures herself, though, that it is only "all 'over the wire'" (47). When the obnoxious impostor confronts her, Nattie realizes that the words she has been transmitting may have actual repercussions in the real world. Both she and the offending intruder deny this possibility, he playfully and she much more vehemently. "It's easier to buzz on the wire than it is to talk, isn't it?" he asks her knowingly, and she replies, "I suppose no one really means what they say on the wire. I am sure *I* do not" (99, 102). Nattie's friends recognize even more clearly than she does the extent to which telegraphy liberates her from intimidation. Nattie explains to Clem, "I—I do not get acquainted quite so easily as Cyn," whereupon Cyn interrupts, "except on the wire!" Nattie tells Clem, "I never talk to strangers," but he too qualifies: "except, perhaps, on the wire" (142, 143). In her communications, Nattie maintains a double standard, following one set of rules in the flesh and another in on-line exchanges.

Thayer subtly suggests a parallel between Nattie's vocation of telegraphy and Cyn's of theater and music, for both involve performances with an ambiguous relationship to real-world events. Whereas Nattie—and, usually, the narrator—describes human relations in terms of wires and switchboards, Cyn sees them in terms of conventional dramas. When Nattie struggles to explain her unhappy relationship with Clem, she sees herself as "the unmeaning instrument of it all." Cyn, however, thinks only, "What a farce it would make!" (212). Once Nattie begins teaching the young singer Morse, Cyn mixes her metaphors, declaring, "a crisis is what is essential to complete the circuit, telegraphically speaking . . . to bring down the curtain on everybody" (198). Like theatrical comedies, Nattie's on-line exchanges involve language that is bandied about for pure pleasure. One's "buzzing over the wire" need not necessarily mean anything other than what the reader chooses to attribute, for telegraphic communications, like fiction, invite the dissociation of signifiers from signifieds.[6]

Just how much meaning incoming telegraphic signals can convey is the central issue of both Thayer's novel and Helmholtz's studies of perception. As the physiologist theorized, arbitrary signs can be used to construct meaning not so much because of what they symbolize as because of their relation to each other. Empirically, one learns to perceive by coming to recognize patterns of neural signs. The operator of a technological nervous system learns the same way, gaining intimate knowledge of her signals without knowing the reality that they represent. When Nattie repeatedly "breaks" in the opening scene because she cannot keep pace with "C's" rapid transmission, he at last responds, "Oh." "For a small one," the narrator remarks, "'Oh!' is a very expressive word" (11). In this two-letter signal, which corresponds more to a sigh than to any meaningful speech, Nattie reads contempt and amusement. As in speech produced by the body, it is not regular symbolism that carries the meaning but the nuances and variations of an individual voice.

Skilled operators were always sensitive to each other's telegraphic styles, and *Wired Love* surpasses the stories of *Lightning Flashes* in showing how much a good telegrapher could read from another's touch on the key (Standage 130). "With the key as with the pen," the narrator explains, "all operators have their own peculiar manner of writing" (92). If all operators were as perceptive as Clem, the double deceit of "Playing with Fire" could never have occurred, for when Nattie tells him she is a "tall young man," he refuses to believe it. "There is a certain difference in the 'sending' of a lady and a gentleman that I have learned to distinguish," he responds (22–23). Nattie retorts that "people who think they know so much are often deceived," whereupon he asks her to "picture, if you like, in place of your sounder, a blonde, fairy-like girl talking to you" (23). At this point a third operator breaks in and tells her not to believe it, but Nattie needs no outside confirmation of her own reading. In Thayer's depiction, the rhythm of dots and dashes conveys gender as unmistakably as voice or handwriting does.

Technological communications, like those sent directly by the body, have a distinct personal style. When Clem "speaks" to Nattie's key instead of to her awaiting ears, an operator somewhere down line "hears" his endearments and pipes up, "That sounds like 'C's' writing!" "My style must be very peculiar," thinks Clem, "to be so readily detected" (173). In the timing of electronic pulses, a good operator could even read the sender's mood. When "C" turns sarcastic in the opening scene and Nattie once again asks him to resume transmission, we are told that "her fingers formed the letters very sharply" (13). Later, when Nattie spurns her lover, thinking he is the vulgar red-haired man, "C" calls her "persistently, savagely, and entreatingly—all of which phases can be expressed in dots and dashes" (111). In

Wired Love, the signs that operators transmit convey both more and less than they desire—more of the sender and less of the reality the sender is trying to describe.

As a sign system with a tenuous relation to the material world, telegraphic language is subject to its own currents and eddies, errors and confluences that create new meanings independent of those its signals were intended to convey. In written and spoken language, miscommunication can occur because of coincidental resemblances between words. By introducing a second level of representation, Morse code offers additional opportunities for misapprehension and punning because of the new accidental similarities it brings. Translated into dots and dashes, "but" and "that" are distinguishable only by the positioning of one pause ("Freaks of the Telegraph" 476). When a message was translated from the sender's handwriting into Morse and then back into written language by the receiving operator—as it was in the late nineteenth century—innumerable errors could arise.

In the first scene, distracted Nattie makes a two-letter substitution, introducing a "fatal" error to "C's" message. A traveler from "C's" town is requesting that someone meet him at the station in Nattie's city, and the original message reads, "Send the horse. . . ." Nattie, however, has transcribed, "Send the hearse. . ." (18). For weeks afterward, "C" teases her with telegraphic puns ("*grave* re-*hearse*-al," "*undertaking* the line"), using the confluences of spoken language to play with its representations over the wire (24). He tells her of an operator who made a similar mistake, in this case by failing to judge a single time interval properly. As the narrator explains, operators leave short spaces between letters and slightly longer ones between words, with the consequence that one can make the appropriate divisions only by knowing a telegrapher's usual style. By misjudging the length of a single pause, this unhappy operator read "John is dead, be at home at three" as "John is dead beat, home at three" (67–68). The sophisticated electronic communications system of 1880 was only as good as its receivers, the ears and brains that analyzed incoming messages. As Helmholtz had shown, these machines could attribute meaning to signs only on a relative basis.

Introducing further noise to the characters' readings of their world are the many fictions that move along their circuits. Nattie and Cyn understand their relationships not just in terms of the messages they receive but in terms of the novels they have read and the theatrical performances they have attended. That fiction sustains them can be seen with comic effect during their beggars' feast. The bread is served on a copy of *Dombey and Son;* Clem eats off of a copy of *Scribner's;* and when Nattie exclaims to Clem,

"I've seen you before!" the artist Norton cries, "that sounds like a novel!" (143). More than any other character, lonely Nattie finds her head full of phrases she has acquired from fiction. She thinks that Norton, who loves Cyn, is looking at the beautiful singer with "his soul in his eyes," all the while conscious that the phrase comes from a "lately-read novel" (75). Even Clem, trying to describe his father's "pig-headed" determination to make him a doctor, compares him to a character in *Our Mutual Friend* (160). Itself a form of fiction, telegraphic communication is always full of other people's stories, and Nattie's society supplies endless texts out of which phrases may be spliced. Betraying his own secret and shaking Nattie's confidence in her lover's wit, the false "C" jokes that on the wire, "all a fellow has to do is to take up a book or a paper, pick things out to say, and go at it without exercising his own brains" (99). Because telegraphic messages have no reliable connection to reality, only a complex metaphorical link, one cannot even be sure whether they consist of original fiction.

Just how much can one know, then, after reading a telegram? How did the telecommunications system of 1880 affect people's understanding of what knowledge was? As Nattie and "C" exchange messages early in the novel, the narrator remarks, "all this time it never occurred to them that . . . they knew really nothing about each other, not even their names" (49). Cyn comments that "it appears telegraph operators have a way of talking together over the wire, knowing little about each other and nothing at all of their mutual personal appearance" (144). But Nattie and "C," as she originally knows him, learn a great deal about each other, if only on a relative basis. As they exchange messages, they come to know each other's styles, moods, fictions, purported feelings, and personalities. As both Nietzsche and Helmholtz claimed, sign systems metaphorically linked to reality do enable a certain kind of knowledge, the only kind of knowledge available to minds housed in living bodies.

When Clem Stanwood reveals that he is "C" and denounces the greasy impostor, Cyn reflects, "we know Clem and 'C,' but Mr. Stanwood is a stranger" (157). Here she acknowledges that one can know another person either through long-distance communications or through another person's descriptions of him. The young people know Clem as their witty dinner partner, but "Mr. Stanwood"—an unknown, formal component of Clem and "C"—eludes them because none of them has yet experienced this aspect of his identity, either in person or over the wires. Nattie posits another identity for "C" when facing the obnoxious usurper, asking herself, "was not 'C,' *her* 'C,' the 'C' whom she knew by his conversation only—'picked out of books!'—an unreal, intangible being, and not this so different person who claimed his identity?" (99). In a revealing moment, Nattie wonders "whether she would be glad or sorry" if "C" were to get an office in her same city (91). "Face to face," she thinks, "we should really be strangers to each other" (60). Sure enough, when "C" does materialize into Clem, her fears that nearness will bring constraint are quickly realized, and she admits that she had more of his company over the wire. As the imagined agent sending titillating telegraphic signals, "C" had a very real and powerful identity, one that could be known and excite hunger for more knowledge.

As indicated in the novel's two-word title, Thayer's novel explores the erotic thrill of uncertain knowledge. When Nattie begins arriving at her office early and leaving it late, she attributes her enthusiasm to the "power of mystery," but her on-line conversations suggest a more fundamental, physiological drive (48). In one of their earliest conversations, "C" asks Nattie to describe herself, suggesting that classic invitation to eroticism, "what are you wearing?" Knowing that she can create images of herself without being fully known gives Nattie pleasure and power. To play the game well, she need only remain consistent in the fictions that she creates.

The secret intimacy and forbidden nature of Clem and Nattie's communications increase their erotic charge. Once the wire has been strung between their boardinghouse bedrooms, they murmur electronically into the "small hours" of the morning, and Cyn and Norton, who begin learning Morse, "hammer away incessantly" (180–81). At last, in a hilarious exchange, Miss Kling exposes the full meaning of her boarders' nocturnal communications. "That any young woman should be so immodest as to establish telegraphic communication between her bed-room and the bed-room of two young men is beyond my comprehension!" she cries. Devilish Clem then responds, "I induced Miss Rogers to allow the wire to come into her room" (245–46, 248). When being wired is forbidden, when it subverts an established power structure, technological connections can bring a physiological thrill.

Despite the excitement of telegraphy, however, and despite all the knowledge and power it promises, the information it provides is as limited as that delivered by any nervous system. Throughout Thayer's novel, characters express their frustration with the restrictions of their communications. In the very first scene, Nattie reflects that "no instrument had yet been invented by which she could see the expression on the face of this operator," and the narrator later remarks, "it was a pity that no telegraphic instrument had yet been invented that could carry the blush on Nattie's cheeks for ["C's"] eyes to see" (13, 79). Like Nattie's own face, which Thayer describes in only the vaguest terms, her communications merely "suggest a possibility" (43). If telegraphy—like the words on Thayer's page—cannot convey expression, then it can convey nothing of Nattie's real appearance, and it makes little sense to describe her face in any more detail.

When one reads *Wired Love*, Thayer implies with her dedication and conclusion in Morse, one is reading a telegram. One is working as an operator any time that one uses one's mind and senses to interpret the world at all. As people longing to connect, people tantalized by opportunities to communicate that reincorporate the restrictions under which they live, all readers are telegraphers, struggling to express themselves and to make sense of unreliable information. Those who felt most like operators, however, were young working women, trapped in low-paying jobs and catching occasional glimpses of more colorful lives.

The Class That Wired Everything: Henry James's "In the Cage" (1898)

Fifteen years after *Wired Love* appeared, Henry James began his own novella about a young, female telegrapher. Unlike Thayer, he wrote for well-to-do readers, and his interests were more epistemological than social. While James and Thayer directed their stories to readers from different socioeconomic levels, they depicted people's struggles to communicate in strikingly similar ways. In each of their stories, telegraphy functions as a nerve net tantalizing the mind with unreliable knowledge. James never had to work long hours for low pay, but he identified with telegraphers and expected his readers to as well. Like Thayer, he challenges claims about the new knowledge and social bonds that supposedly result from instantaneous electronic communication.

By the 1890s, the European and American public had become so dependent on telegraphy for long-distance communication that threats of strikes by telegraphers produced serious social concern. In the 1870s, Great Britain nationalized its telegraph network, incorporating it into the postal system, and the subsequent drop in price created an enormous increase in "traffic" (Moody 55).[7] In 1897, London had over three hundred offices open from 8:00 A.M. until 8:00 P.M. That year, in London alone, 2,214 female telegraphers (about a third of the total number) helped to dispatch 27.5 million telegrams (W. Stone 244). Despite forty-eight-hour work weeks and extremely low pay, there were many more job candidates than positions available, and hungry job-seekers—usually under the age of eighteen—

could assume their "dot and dashing labors" only after a competitive exam and a five-month training course (W. Stone 245).

During the summer of 1895, when James was starting to write "In the Cage," British telegraphers' anger became acute, and a committee headed by Lord Tweedlemouth began to look into their grievances. Female workers, who were especially vociferous, protested about having to work the 2:00 P.M. to 10:00 P.M. shift and stated that they could not survive in London on their meager pay (Moody 55–56). Unconvinced after two years of hearings, the Tweedlemouth Committee decided to lower telegraphers' salaries to the level of postal sorters', and in the summer of 1897, both male and female workers threatened to strike. This prospect frightened the public, for two such strikes had already occurred in the United States. While James was writing "In the Cage," the *London Times* printed daily reports on the telegraphers' negotiations, articles that he could not possibly have missed (Moody 55–57; W. Stone 244–45).

James's story, however, is much more than a second-hand report on workers' discontent, and his own descriptions of his inspiration indicate that he had been thinking about telegraphy for some time. As a member of the class that "wired everything," James spent a good deal of time in London's telegraph offices, and he remarked that the local office had "so much of London to give out, so much of its huge perpetual story to tell" (Moody 57). In the early fall of 1898, James moved to Lamb House in the Sussex countryside and was "reduced . . . to communication with London by letter and telegram; indeed, the telegram began to play a new role in his life" (Edel 470). Author of innumerable messages, he knew personally how much information passed through telegraphers' hands, and he considered "what it might 'mean' . . . for confined and cramped and yet considerably tutored young officials of either sex to be made so free, intellectually, of a range of experience otherwise quite closed to them" (W. Stone 243). What fictions might arise in the mind of an impoverished telegrapher who spent forty-eight hours a week transmitting messages about unreachable luxuries? How would a person respond, imaginatively, to this Tantalus torture, and just how much could a person know about this life that eluded one's grasp?[8]

The public of the 1890s knew that resentful employees gave poor service, but in the case of angry telegraphers, there was much more to fear. As shown in *Wired Love*, telegraphy aroused very reasonable concerns about privacy, which could be maintained only as long as workers agreed to maintain it. "Blackmail," observed Alexander Welsh, "is made easier by communication technology. . . . Blackmail is an opportunity afforded to everyone by communication of knowledge at a distance" (Moody 53). In the United States, telegraphers leaked information about troop movements during the Civil War. Western Union once sold election news, and there were numerous insider-trading scandals with telegraphers as sources (Moody 57–58). In Great Britain, such momentous leaks had not yet occurred, but no one knew how far dissatisfied workers could be trusted. It seemed inevitable that workers who spent their lives arranging other people's amusements would someday use their customers' information to their own advantage.

Henry James's novella "In the Cage" depicts people's dissatisfaction with telegraphy on many levels. Told from the perspective of a young, female telegrapher, it conveys her anger over the socioeconomic distance between the rich, who "wired everything, even their expensive feelings," and the poor, who transmitted their messages (James 184). Like Nattie, James's protagonist is in her late teens and is living a working-class life despite her middle-class background. Like Thayer's character, too, James's telegrapher is intelligent and hungry, longing for experiences foreign to her cramped existence. James's story, however, goes far beyond class grievances in representing the ironies of telegraphy. Like stories by telegraphers, it is an epistemological study exploring what one can and cannot learn through long-distance communications. As in *Wired Love,* this knowledge proves much more scanty than one would hope. According to James's biographer Leon Edel, "In the Cage" is "of a piece with [James's] immediate feeling of being cut off from London." Once the writer moved to Sussex in 1898, his "sense of being confined and 'out of things' . . . contributed to his imagining a young girl confined daily to a little cage" (475–76). Isolated not just from London but from his native United States as it battled Spain, James wrote to his brother William, "I am near enough here to hate [the war], without being, as you are, near enough in some degree, perhaps, to understand" (James and James 351). At the time James was completing his story, he was living, like his protagonist, through long-distance communications, and he knew all too well how little knowledge they conveyed.

For James, telegrams offered statements out of context that never revealed the true essence of a person or situation. In his view, however, even face-to-face contact did not guarantee real knowledge of another human being, nor did it ensure successful communication.[9] Believing that "relations stop nowhere" and viewing society, as did Eliot, as an ever-changing web, James regarded all human relations as fragile and tenuous (Wicke, "Henry James's Second Wave" 146). As Walter F. Wright has proposed, James's use of telegraphy in the story reminds readers that "one does not really know the full thoughts of others; what he has to work with is symbolized by the abridged and cryptic telegrams—fragments which can be arranged in patterns only by an effort of the imagination" (175).[10] With their enforced abbreviations and their mutilated expressions of human feeling, telegrams could serve as a metaphor for human communications in general.[11]

James opens his story by telling his readers, "It had occurred to [the telegrapher] early that in her position—that of a young person spending, in framed and wired confinement, the life of a guinea-pig or a magpie—she should know a great many persons without their recognizing the acquaintance" (174). From the first sentence, he makes it clear that the woman's knowledge, a problematic knowledge empty of acquaintance, depends on her "position."[12] As James has constructed his sentence, her perceptions of this position are confined within dashes, framed by the essential realities of position and knowledge. It is because of her "position" that James has a story to tell, and his initial portrait of the cage is as ironic as the tale itself.[13]

Working behind a wire-mesh screen that separates her from her customers and the surrounding grocery store, the young telegrapher feels like two different animals. As a guinea pig, she is a passive object manipulated by experimenters outside her wire box, but as a thieving magpie, a surreptitious collector of glittering objects, she is the active experimenter. The function of the cage, James implies, depends entirely upon one's perspective: "this transparent screen fenced out or fenced in, according to the side of the narrow counter on which the human lot was cast" (174). As the story's central symbol, the cage suggests its essential questions in a very real and physical way. How much can one know? Should one live as a guinea pig, passively yielding to social restrictions, or as a magpie, stealing information that may lead to a better life? Both confining and liberating the telegrapher, her cage acts as an interface in the social, sexual, and epistemological play between inside and out.

James's initial description of the cage immediately suggests the maddening divisions of social class. Like Thayer's Nattie, James's protagonist is the "victim of reverses," although she is much angrier about the loss of her social position (177). In James's story, the telegrapher's father is dead, her brother "lost," her sister "starved," and she lives alone with her alcoholic mother (187). Forced to work, she spends her life counting her wealthy customers' words and calculating the prices of their communications, pausing only to read ha'penny novels during her forty-minute lunch breaks. Although she hates her job and her rich customers, she still considers herself superior to barmaids who "band[y] slang," convinced that her perceptiveness and intelligence set her apart from women born into the working class (204).

In her cage, she experiences the telegrapher's Tantalus torture, transmitting endless messages about a life from which she is forever barred:

> What twisted the knife in her vitals was the way the profligate rich scattered about them, in extravagant chatter over their extravagant pleasures and sins, an amount of money that would have held the stricken household of her frightened childhood, her poor pinched mother and tormented father and lost brother and starved sister, together for a lifetime. (186–87)

Confined by her wire, she rages against the "immense disparity . . . from class to class," reading signs of a life that she can never enter.[14] In August, she thinks resentfully, there "were holidays for almost every one but the animals in the cage" (213). She herself has not left London in twelve years.[15]

From her cage, she can see through these barriers, converse through them, and conduct all sorts of business, but they define her "position" even as she observes life on the other side. When forced to consult her handsome customer Everard because he "formed some of his letters with a queerness," she and the aristocratic Captain end up "bringing their heads together over it as far as was possible to heads on different sides of a wire fence" (206). While their common penchant for queer letters draws them together, the transparent barrier separating them remains firmly in place.

James implies, however, that even members of the upper classes are confined by such restrictive gratings. The attractive Everard is "clenched in a situation" and in the end is effectively "nailed" by his aristocratic lover (243, 265). As the novella concludes, none of its three impending marriages promises to be happy—neither the protagonist's, to the grocer Mudge; nor Everard's, to Lady Bradeen; nor that of the protagonist's confidante Mrs. Jordan, to Lady Bradeen's butler (Salzberg 63). As Stuart Hutchinson asks, in this novella of communications, "who isn't in some sort of cage?" (20).

Yet even as the "wired confinement" of her business separates the young woman from the life she desires, it protects her from danger. After she and Everard meet face to face and he expresses sexual interest in her, the telegrapher realizes that "to be in the cage had suddenly become her safety, and she was literally afraid of the alternate self who might be waiting outside" (240–41). The cage shields her not just from upper-class men's advances but from her own weary desires to accept them. Her alternate, sexual self, she realizes, might actually engage in the affair that she has been imagining. To live it in her mind is one matter, but if she were to act on it, as Everard seems to wish, she could lose what little control she has over her life. Imagining that Everard regards her with "supplicating eyes and a fever in his blood," she returns his look, "supplicating back, through the bars of the cage"

(263). She is all too glad, however, that these bars are in place. Fleeing to her professional role, she torments him as any petty official might, so that the cage that screens her off from him lets her retain some power.

In reality, the protagonist will soon be leaving her wire cage, for she is engaged to marry a grocer, Mr. Mudge. She is deferring this fate as long as she can, for it will forever bar her from the upper class that she loves and hates. While marriage will free her from menial labor, it promises work just as meaningless and boring. The telegrapher's struggle for power involves her sexuality as much as her financial position, and while she spends her days accepting words "thrust . . . through the gap," she retains some level of autonomy as long as she remains single and childless (174).[16] Marriage, she suspects, will be just another cage, another screen between her and the life she desires.

Despite the woman's perception of an "immense disparity" between classes, the lives of London's workers and its leisure class were closely interwoven. Like Eliot, James offers numerous images of threads, knitted or woven together, to describe the structure of society. Mudge, whose perceptiveness exceeds the woman's estimation, envisions a healthy economy as a web of relations (Salzberg 64). In his grocer's eyes, "the exuberance of the aristocracy was the advantage of trade, and everything was knit together in a richness of pattern that it was good to follow with one's finger-tips" (James 202). She finds this web infuriating, however, for despite the elaborate connections, wealth and power move only along certain lines.

With her mind and body, the telegrapher operates as a key cross-point in this social web. As Jennifer Wicke points out, in the late nineteenth century, "women came to mediate exchange. Communication flow[ed] through them, telegraphically or otherwise enhanced; information traveling across class lines collocate[d] in them; the mechanisms of mass cultural transfer of libidinal, commodity desire [we]re set up with 'woman' at the switch point" ("Henry James's Second Wave" 148). In the midst of all this traffic, James's telegrapher has a "gift," or so she thinks, for "keeping the clues and finding her way in the tangle" (James 188). Like a lacemaker, she believes, she can keep innumerable different strands distinct in her memory.[17] When one considers her work, the wire mesh of the protagonist's cage becomes less a barrier than an interface; her workspace is "permeable, a nexus as much as a site" ("Henry James's Second Wave" 146). In James's story, the telegraph that annihilates distances brings haves and have-nots into close proximity with each exchange of thoughts among the wealthy.

This maddening nearness to forbidden pleasures makes the young telegrapher all the more hungry for power. Like *Wired Love*, James's telegraphic tale is a story of hunger, depicting a telegrapher who craves recognition and appreciation. Finding none, she begins to enter her customers' lives, maneuvering to obtain any kind of control over them that she can.[18] Although she may not be as intelligent as she thinks she is, she is able to penetrate the coded messages Captain Everard is sending to his aristocratic lover. Jealous of her friend Mrs. Jordan, who enters wealthy clients' homes to provide them with flowers, the telegrapher brags about her knowledge of her well-to-do customers' lives. In a pathetic exchange, the two marginalized, déclassé women compete to show their minimal influence over the rich. "I doubt if you 'do' them as much as I!" exclaims the telegrapher. "Their affairs, their appointments and arrangements, their little games and secrets and vices—those things all pass before me." "I see. You do 'have' them," replies Mrs. Jordan ironically (196). To Mudge the telegrapher boasts, "I've got them all in my pocket," but he is clearly no more impressed than her friend (232). Unable to achieve any significant power or control over others, she seeks it, increasingly, in her fantasies.

Gradually, the young telegrapher convinces herself that she is indispensable to her attractive customer Everard. She imagines that she has mastered the code through which he communicates, and she even suspects that he has developed feelings for his Hermes, the "extraordinary little person" in the telegraph office (211). First mentally, then bodily, she leaps out of her cage, correcting Lady Bradeen as the wealthy woman struggles to formulate a message (213). Stalking Everard, the telegrapher enjoys a long conversation with him in a park and pushes him to acknowledge her role in his life. "I'd do anything for you," she tells him, offering her services (223). The flustered Everard seems a bit too interested in her offer, however, and when he begins dallying around the office, she retreats to her official role. Seeking power in another, more professional way, she toys with him and prolongs his suspense when he asks her help reconstructing a telegram that has been intercepted by Lady Bradeen's husband. In the end, her quest for control and recognition brings her right back into the cage. Finding no appreciation in her work, she seeks it in her imagination; achieving none through her fantasies, she seeks it again in her work.

"In the Cage" is thus a tale of disillusionment and humiliation, but its conclusion involves far more than the protagonist's pitiful failure to increase her power.[19] Her inability to influence others derives from her failure to know. Even as a cross-point in a network vibrating with information, she has learned next to nothing. In a final exchange with Mrs. Jordan—who is about to marry Lady Bradeen's new butler, Drake—the telegrapher discovers that Everard has large debts and is being forced to marry the "compromised" Lady Bradeen. By altering his lover's telegram incorrectly, she may actually have caused its interception (Norrman 427). Even this act would have altered his life minimally, however, since he had little desire to marry his pregnant lover in the first place. Lady Bradeen has "just nailed him," having held the real power all along (James 265).

Humiliated, the young telegrapher vows to marry Mudge as quickly as possible, concluding that "reality . . . could only be ugliness and obscurity" (261). Her pride in her intelligence, however, is never quite eclipsed, for in the story's final line, she thinks how strange it is that "such a matter should be at last settled for her by Mr. Drake" (266). Although not fully crushed, she resigns herself to socially imposed restrictions on her power and knowledge. Scared straight, she is shaken not just by this new vision of overwhelming social forces but by the obvious inaccuracy of her own readings (Bauer and Lakritz 69). In the future, one suspects, she will seek power in more socially sanctioned ways, as a wife and mother influencing her children and her grocer husband.

While it seems questionable to read "In the Cage" as a "morality tale," the story can certainly be read as upper-class wish fulfillment (Moody 55). In James's story, the class that "wired everything" encounters an intelligent, unscrupulous telegrapher who tries her best to pry into a customer's affairs but is unable to discover anything. Taken at face value, "In the Cage" tells the rich not to worry and the poor not to bother interfering with their lives. Knowledge, implies James, may be obtained in ways one does not expect, but it will never be obtained by trying to decode people's long-distance communications.

While the cage in James's tale suggests the divisions of class and sexual power, it also represents the restrictions and powers of consciousness. No simple box, the cage has quite a complex structure. At its center lies the sounder, "the innermost cell of captivity, a cage within the cage, fenced off from the rest by a frame of ground glass" (180). Like a sensory ganglion, or perhaps a small brain, the protagonist's communications center processes information and passes it on. In her work, the telegrapher operates as an interneuron in this brain, reading and translating messages but lacking any direct access to the outside world. Just as the brain sacrifices its own sensitivity to maximize its integrative powers, this mechanized communications system keeps its army of workers from knowing the real situations to which their information refers. A telegraphic network is not conscious, of course, but both the nervous system and technological communications systems rely on specialized cells that lack direct access to the world but excel at reading one another's inputs. The great irony of telegraphy is that its workers must sacrifice direct contact in order for the complex communications network to function.

As an interneuron, the telegrapher has access only to language, to other cells' "signs" of the world. Significantly, this telegrapher differs from Nattie in that she does not operate the sounder herself—she is never physically connected to a network of wires. Instead, her link to the communications web is forged through language, as she encodes and tallies up her customers' words. Even when she perceives the web as a continuation of her nerves, she is being inspired more by shared codes than by continuity with an apparatus (Hayles, "Escape and Constraint" 9, 5). Because James shows how poorly language represents actual feelings and events, the woman's position in the cage suggests the fate of all people who rely on language for knowledge.[20] The cage's capsule within a capsule suggests the structure of the telegrapher's own mind, in which hunger and anger are repressed so that she may maximize her control over others.[21] Carren Kaston speculates that "consciousness itself may be [the] cage from which character needs to be rescued" (Gabler-Hover 262).

In James's descriptions, the telegrapher's cage and her activity in it resemble the structure and functions of a living body. Like the protagonist's own corpus, the "frail structure of wood and wire" is permeable and is breached each time someone "poke[s] in a telegram" (175, 245). As she works, the telegrapher feels like a brain in its organic housing, "a changing pushing cluster, with every one to mind at once" (183). With the demands made upon her own mind and body, she identifies herself physically with the work she performs. Like other workers subjected to Taylorist principles, she knows that she is seen as a machine, and she deeply resents it. To Everard, she expresses her anger toward people who treat her "as if I had no more feeling than a letter-box." To console her, he replies, "Oh, they don't know!" but she snaps, "don't know that I'm not stupid? No, how should they?" (226).[22] What she gets out of her work, she tells him, is the "harmless pleasure of knowing," but even this pleasure is elusive, and she rages against her "harmless" status (227). What good is knowing, she wonders, if one can never act on what one knows?

If telegraphers were regarded—and saw themselves—as extensions of their machines, there were also Victorians who saw social communications networks as alive. As Richard Menke proposes, telegraphy promised to "relocate consciousness in the electrical pulses of the network" ("Telegraphic Realism" 14). According to Andrew Moody, many people of the late nineteenth century "believed that the Post Office was alive and possessed intelligence." Debates broke out between those who wanted a "mechanical" communications system, which offered absolute privacy, and those who wanted a "living" one, in which workers had "personal relations with—or at least personal knowledge of—the public" (58–59). At first, James's telegrapher prefers the "living" model, finding it dehumanizing to be seen as a "letter-box." Once she sees what personal relations may involve, however, she is all too happy to return to the mechanized system.

Like Eliot, James uses the term "sympathy," associating social bonds with physiological ones. In the story of the telegrapher, these allusions to sympathetic links tie physiological and emotional connections to those of the telegraphic network. On a primary level, sympathy refers to the protagonist's own mind and body. She has "wonderful nerves," we are told, and is "subject . . . to sudden flickers of antipathy and sympathy" (177). Resisting the temptation to warn her female customers about their affairs, she thinks about "the hazard of personal sympathy," a "harmless" emotion as long as it is confined to the cage of her own mind (188). The telegrapher even imagines that with Everard she lives in "a kind of heroism of sympathy," and she is frustrated that she cannot show him the strength of their bond (205). When she tells Mrs. Jordan how much she hates the rich, her friend proclaims, "Ah, that's because you've no sympathy!" James then writes that "the girl gave an ironic laugh, only retorting that nobody could have any who had to count all day all the words in the dictionary" (197). Forced to break language into its component particles, the telegrapher is all too aware of her inability to form true personal, emotional bonds. While she denies it through most of the story, she realizes that in her cage, she can know no one and nothing at all.

Like Nattie Rogers and the characters in the telegraphers' stories, James's protagonist cannot see what is happening at the other end of the wire. Her "knowledge," of which she is so proud, consists solely of guesses, and she proves to be a poor detective.[23] Because her inputs are limited, she must always make do with clues. In October, for instance, "she made out even from the cage that it was a charming golden day" because "a patch of hazy autumn sunlight lay across the sanded floor" (235). When Everard's wealthy lover first enters her office, "it floated to her through the bars of the cage that this at last was the high reality" of her dreams (181). In almost every case, what she perceives as coming from outside of her cage is projected from the inside out. Although Cocker's, the grocery store in which she works, receives no telegrams through its sounder, the protagonist receives a great variety of handwritten messages to be processed for transmission. All of her "knowledge" of Everard can be traced to her overzealous readings of these ambiguous signs. "Every time he handed in a telegram," she thinks smugly, "it was an addition to her knowledge: what did his constant smile mean to mark if it didn't mean to mark that?" (205). Even as she formulates the thought she is reading into his writing, drawing unfounded conclusions from the punctuation of his face. For the telegrapher, knowledge is something that one *makes*, although she never perceives it in this way.

Throughout his story, James encourages readers to ask what "knowledge" means.[24] "The correspondence of people she didn't know was one thing," the telegrapher thinks to herself, "but the correspondence of people she did had an aspect of its own for her even when she couldn't understand it" (194). If one reads this sentence in the context of James's first one ("she should know a great many persons without their recognizing the acquaintance"), one wonders of what her knowledge consists. She comforts herself when she is unable to decipher Everard's coded telegram—at least she knows him, she thinks—but what does she know? Her only source of information has been his previous telegrams, which have been equally unintelligible.

Like Eliot, James's telegrapher believes that knowing means drawing connections. On vacation with Mudge, who always wants to get his money's worth, she realizes that "she got back her money by seeing many things, the things of the past year, fall together and connect themselves, undergo the happy relegation that transforms melancholy and misery, passion and effort, into experience and knowledge" (230). Eliot would never approve of the connections the telegrapher draws, however, for they have no roots in personal experience or human sympathy. While the telegrapher exults, "How much I know—how much I know!" believing that she is better informed about Everard's life than is Lady Bradeen, the epistemological connections she is making have value only in her own mind. The secrets that she carries in her "small retentive brain" are fictions she has spun herself (212, 187). In the end, Mrs. Jordan is right; the telegrapher has no sympathy, for her knowledge—based on neither acquaintance nor fact—has no basis in real human relations.

In James's story, the telegraph allows one to transmit information and act at a distance, but it never allows one to know. The protagonist, he suggests, is not the only one living by telegraph. All those who wire their "expensive feelings" believe that they are staying in touch by transmitting phrases with a fixed cash value. Her wealthy customers' telegraphic "talk," the woman reflects, is "so profuse sometimes that she wondered what was left for their real meetings" (184). For her own sake, she rejoices that Everard is "leading [his life] so much by telegraph," for as he maintains his relations with others, every exchange reinforces her perceived relation to him

(207). When he and Lady Bradeen wire each other, the protagonist fancies that "she [is] with the absent through her ladyship and with her ladyship through the absent" (211). In reality, she and Everard are playing the same reading game, accepting fragmented signs for actual presence.

At times the telegrapher expresses herself as though her overzealous reading were actual observation, thinking, "she would defy any other girl to follow him as she followed" (206). Once she has met Everard outside of the cage, however, she seems well aware of the difference between imagination and reality. "It was as if they had met for all time," she reflects, and "it exerted on their being in presence again an influence so prodigious" (236). Reality has jabbed into her imaginative web, and she does not like the intrusion. Preferring to deal with Everard from within her cage, she retreats into her imagination.[25]

In designing his tale, James intentionally set his story in the telegrapher's own mind, explaining that "the action of the drama is simply the girl's 'subjective' adventure" (Norrman 425). Once he has described her "position," he tells readers that "she had surrendered herself . . . of late to a certain expansion of her consciousness" (James 178). He then invites them into her cage so that they can experience this expansion with her. Even as readers pity the woman's impotence, they share her fate. When she reconstructs Everard's romance and then reads herself into it, she makes the same faulty assumptions that readers do, believing that from their limited perspectives they can acquire real knowledge. Although her case is extreme, James hints that we are all telegraphers, relying on fallible information systems to construct trustworthy knowledge. Much to the relief of the class that "wired everything," it was simply not possible to "understand the private details of another's life . . . without a personal relation to that person" (Moody 64). Like an organic communications system, James's tale implies, a technological one can provide only limited information about events outside of the cage.

Because the story is set in the protagonist's own consciousness, James can expose her misreadings only by contrast. This he does very effectively with reports from Mrs. Jordan. While the florist's information is second- or third-hand, it is based on personal acquaintance and, originally, on actual observation. It has never been fragmented or encoded, and it is not a text waiting to be read. When readers encounter James's descriptions of Mrs. Jordan's activities, they instantly believe this eyewitness account, and it appears that he wants them to believe it. Because of her work, Mrs. Jordan is constantly "in and out of people's houses" (James 177). "There was not a house of the great kind," describes the narrator, "in which she was not . . . all over the place" (192). To emphasize Mrs. Jordan's more intimate contact, James describes it in physical, organic terms: "Mrs. Jordan . . . touched [Everard] through Mr. Drake, who reached him through Lady Bradeen" (263). A long-distance correspondent to the protagonist's encaged mind, Mrs. Jordan offers real information, and the telegrapher is deeply jealous, humiliated by her friend's superior access. "It was better surely not to learn things at all," she thinks disgustedly, "than to learn them by the butler" (259). Like other Victorians, she has far more respect for telegraphy than for gossip, but she is forced to admit that personal relations are a superior source of knowledge.

James offers readers this unpleasant possibility by exposing the telegrapher's mistaken guesses. She is "wrong about everything," and her disillusionment is all the more devastating because of her earlier confidence in her knowledge (Aswell 377; Norrman 425). She is wrong, for instance, in her conviction that Everard is not the Captain's real name and in her intuition that she will never see him again. More significantly, she errs in believing that he wants to marry Lady Bradeen and in the impossibly remote prospect that Mrs. Jordan will marry Lord Rye (Norrman 425). The telegrapher's most pitiful misreading, however, is her conclusion that Everard loves her and has begun writing his messages especially for her. Poring over one of his telegrams, she thinks, "The point was not that she should transmit it. The point was just that she should see it" (242). The text of this telegram—"absolutely impossible"—sums up the world's reply to her conjecture.

When the telegrapher crosses over into reality, she almost convinces the reader that she possesses real knowledge. Seizing control when Lady Bradeen falters over the code, the protagonist suddenly volunteers, "isn't it Cooper's?" Describing the telegrapher's rush of power, the narrator writes, "it was as if she had bodily leaped—cleared the top of the cage and alighted on her interlocutress" (213). Pouncing on her aristocratic customer, the telegraphic tiger believes she has cleared the barrier between signs and reality. Sadly, though, she has erred again, not just in her knowledge but in her judgment. By changing the wrong sign, she may have caused Everard to be discovered, and by humiliating Lady Bradeen, she has permanently lost a customer (Norrman 427). On all levels, her attempt to act from within the cage proves disastrous. Despite the sheer quantity of information she encounters, she is effective only as a processor, never as a motor neuron.

As a reader of signs, James's protagonist seeks power through language; she tries to know and control others by controlling their words (Aswell 376). As she pores over Everard's messages, she struggles to crack a code within a code; she knows Morse and English but is challenged by the symbols of the aristocratic lovers. "*His* words were mere numbers," she perceives. "They told her nothing whatever; and after he had gone she was in possession of no name, of no address, of no meaning, of nothing but a vague sweet sound and an immense impression" (182). Despite her exultant claims of knowledge, she never deciphers this "hieroglyph" (Wicke, "Henry James's Second Wave" 149).[26] Like language itself, the remote lovers' code defies all control. It slips through her fingers and her eager mind, but it is never hers to wield, hers to master.

Like Eliot, James presents language as a vital connecting medium, one that links person to person like the telegraphic network along which it is transmitted. In late nineteenth-century London, language both separated and connected. In British Victorian society, one's use of words determined one's social class, yet at the same time, language constituted an evolving, dynamic system in which people separated by an "immense disparity" were constantly imitating one another. Criticizing the rich to Mrs. Jordan, the young telegrapher consciously employs a phrase she has heard her wealthy customers use: "Then, as she had heard Mrs. Jordan say, and as the ladies at Cockers even sometimes wired, 'It's quite too dreadful!' " (202). Like her ambitious friend, the telegrapher hopes that by imitating the language of the upper classes, she may acquire some of their privileges. In James's novella, as in London society, language is power. In her cage, the woman can wield it very little, but she sits in the heart of its field, with its energy sizzling all around her.

As she "count[s] words as numberless as the sands of the sea," the protagonist spends her days translating phrases into pounds and pence (174). At Cocker's, language means money, which is likewise controlled and scattered by the upper classes (Wicke, "Henry James's Second Wave" 147). Supposedly, words are mere signifiers, distinct from the real objects they represent, and the telegrapher relies on the freedom of language to escape reality through her imagination. The prospect that words are things in themselves, however, with a distinct monetary value, undermines her attempts to win power through language (Bauer and Lakritz 68) . Even in the ha'penny novels that supply her with images, words have a price. Living in a capitalist society in which almost anything can be assigned a monetary equivalent, she thinks even of her moments with Everard as a "little hoard of gold in her lap" (220).[27] In his story, James plays with an economy in which language, money, and power are interchangeable. While his telegrapher inhabits the realm of the equals sign, she controls none of the three.

To complicate these exchanges, James introduces one more term to this cruel mathematics: that of organic tissue, represented ironically by Mrs. Jordan's flowers. Flowers, like words, can be used to express ideas and feelings.

As Mrs. Jordan knows, there is a whole code to floral arrangements, and a bouquet, with its carefully selected elements, can be a message more informative than a telegram. Through the conversations between the two déclassé women, James reveals that in their society, flowers mean power, just like money and words. Certainly, his two hungry characters think so. Mrs. Jordan uses images of her own long-lost garden at the vicarage to remind herself and others of her former middle-class status. "This small domain," explains the narrator, "bloomed in Mrs. Jordan's discourse like a new Eden, and she converted the past into a bank of violets by the tone in which she said, 'Of course you always knew my one passion!'" (190). The protagonist compares her dealings in words to Mrs. Jordan's dealings in blossoms, reflecting that "a thousand tulips at a shilling clearly took one further than a thousand words at a penny" (193). Frequently ordered by telegram, flowers—like words—could be sent for a fee, and the ability to scatter them indicated one's wealth and influence.

Images of flowers pervade James's story, not just in its descriptions but in its very concept. Describing how he got the idea for his novella, James wrote that in local telegraph offices, "some such experimentally-figured situation as that of 'In the Cage' must again and again have flowered" (Stone 243). Within the story, blossoming flowers represent the characters' thoughts and feelings. "Preparation and precaution," the narrator tells us, are "the natural flowers of Mr. Mudge's mind" (229). While Mr. Mudge's blossoms seem well nourished and not particularly attractive, Everard's are delicate and fragile. The telegrapher recalls Everard's inarticulate "Only I say—see here!" as "the sweetest faintest flower of all," trying desperately to suck up its barely perceptible perfume (245). Describing the protagonist's Tantalus torture, the narrator tells us, "the nose of this observer was brushed by the bouquet, yet she could never really pluck even a daisy" (186). Like Mrs. Jordan, she must spend her life smelling the blooming thoughts of others, struggling vainly to control the process of their arrangement.

As floral and textual arrangers, the women hold analogous jobs, a point that James stresses in all of their exchanges. Mrs. Jordan tells the telegrapher:

> "You should see Lord Rye's."
> "His flowers?"
> "Yes, and his letters. . . . You should see his diagrams!" (195)

Through such juxtapositions, the reader comes to see words as living and blooming, at least for a time, objects that can be bought and sold and that have been cut off from their original roots.

Both inside and outside of her cage, James's telegrapher encounters rootless words, signifiers with no clear relation to any known reality (Eros). If she misreads Everard's coded, oddly lettered messages, she fares even worse—as does he—with his spoken words when the two talk face to face. As they converse in the park, they make vague, open-ended statements, revealing little of the feelings they may or may not have. Presuming that Everard thinks as much about her as she does about him, the telegrapher tells him, "anything you may have thought is perfectly true" (220). Everard claims that he has "thought a tremendous lot," but he never specifies his thoughts and is most likely just trying to please her. When at last he "appear[s] on the point of making some tremendous statement," he just "let[s] it altogether fall" (225). His final, pitiful appeal to her—"see here— see here!"—could mean almost anything, and it is for this reason that she values his words (228). Directed, at last, to her alone, his vague plea for her to stay could be read as a cry of love, but it is probably nothing more than a bored man's request for female company. As in Eliot's novel of linguistic and moral bonds, the language that ties people together fails to communicate their true feelings.

Toying with Mudge after her encounter with Everard, the telegrapher describes her relationship with the Captain in especially vague terms. To play on his emotions, she reveals to her fiancé how she told Everard she would "do anything for him." "What do you mean by 'anything?'" he asks, and she answers, "Everything" (233). Mudge's response—silently pulling out a bag of chocolate creams—is just as informative as her confessions. When the grocer utters the meaningless phrase, "I say," she thinks "this was an ejaculation used also by Captain Everard, but oh with what a different sound!" (233). Like Thayer's narrator, she implies that the sound, not the words, carries much of the meaning, revealing the unreliability of language for communication. When she remembers Everard's words, it is really their ring that has registered. The narrator reports, "'See here—see here!'—the sound of these two words had been with her perpetually" (237). Because she has seized on their sound rather than their sense—if they ever had any—she can give to them whatever meaning she chooses.

Faced with fragments and indecipherable symbols, James's telegrapher spends her life reading in (Bauer and Lakritz 62). Bored by the task of breaking language into its component particles, she reconstructs it in her imagination even as she decomposes it in fact. In a description strikingly reminiscent of Eliot's Dorothea, James asks, "what may not . . . take place in the quickened muffled perception of a young person with an ardent soul?" (210). Like Eliot's character, the telegrapher sees both too well and too poorly.[28] In deciphering messages, the narrator tells us, "sometimes she put in too much—too much of her own sense; sometimes she put in too little" (180). Guessing the referents of her customers' words is a challenging task, for she never has more than half of the story. While she spends her life counting the words of Everard's telegrams, she never has access to the replies.

Thinking critically about her task of creative reading, James's telegrapher mixes her metaphors:

> She read into the immensity of their intercourse stories and meanings without end. . . . Most of the elements swam straight away, lost themselves in the bottomless common, and by so doing really kept the page clear. On the clearness therefore what she did retain stood sharply out; she nipped and caught it, turned it over and interwove it. (189–90)

In her cage, the telegrapher reads actively and selectively. Feeling like a powerless animal, she "nips" and "catches" the linguistic food that powerful customers toss, but as a thinking, imaginative human being, she takes the initiative, turns it over and studies it, and begins constructing a pattern.[29] Behind her wire, she is both powerful and powerless, living out the fate of all human beings who can communicate only with slippery words and only from within the cages of their bodies.

At times, writes James, the telegrapher feels that "all the wires in the country seemed to start from the little hole-and-corner where she plied for a livelihood" (187). Because of her endless task of preparing information for transmission, she envisions herself as the center of a vast communications net. As Eliot's imagery implies, however, one must take great care in assessing one's true "position." Because each individual is the focus of innumerable relations, each is a nexus of sorts, but fatal errors will be committed if each sees him or herself as the one true center of an immense web. The telegrapher's work makes this vision of herself more reasonable than self-centered Rosamond's. As business offices and communications networks increasingly employed women, young typists and telegraphers rightly perceived their minds and bodies as interfaces in the rapid flow of information (Wicke, "Henry James's Second Wave" 148).

A parasite in the telegraphic net, James's telegrapher begins to interfere with the messages she makes possible. Tracking her customers "mainly in one relation," she profits from a long-distance relationship that requires her presence (James 189). With the noise that she creates in the wires, she may even have transformed the relation she is studying.[30] A "psychic vampire" of sorts, an intermediary in an essential chain of communication, James's telegrapher tries to take control of others, "steal[ing] other people's vitality without even touching them" (Gabler-Hover 266–67). Where people must rely on communications networks, the author suggests, they will always have to deal with parasites. One will hear frustratingly little over noisy lines, and one must be wary of what one hears and of any mind that helps to transmit it.

Chapter 6
A Web without Wires

As the nineteenth century drew to a close, improvements in communications technology only encouraged comparisons between living and manmade systems. Just as telegraph operators of the 1870s looked on their wires as a living network, creative writers who conducted their correspondence over the wires reflected on the similarities between their nations' and their bodies' transmissions of thoughts. The use of the telephone, from 1877 onward, further stimulated people's imagination. It seemed like a natural progression: if writing and speech could be transmitted as electrical signals, why should thoughts not someday be transmitted directly?

In the popular mind, both nervous and electrical impulses consisted of waves conducted through the ether. A hundred years previously, the telegraph and the telephone would have been dismissed as fantasies, and if the efficiency of communications technology continued to improve, it seemed reasonable to expect that networks would someday transmit thoughts themselves. Widespread mesmeric demonstrations reinforced people's belief that one mind could influence another, and the reports of spiritualists suggested that minds could communicate not just across space but across time.

In 1876, British scientist William Barrett (1844–1925) presented his studies of thought transmission to the Royal Society, asking that this unexplored mode of communication be examined scientifically. Receiving no serious response, he wrote to the *London Times,* requesting that other people who had witnessed thought reading compare their experiences to his. Barrett was overwhelmed with letters, for by the late 1870s both scientists and laypeople had begun to suspect that minds could transmit thoughts without the primitive use of wires.

Mental Telepathy: The Society for Psychical Research

Having received only "derision and denunciation" from England's official scientific society, Barrett decided to found one of his own (Barrett 53). In

January 1882, with help from F. W. H. Myers (1843–1901) and George John Romanes (1848–94), Barrett organized the first meeting of the Society for Psychical Research. In an era in which quantitative, "scientific" methods were being extended to social and medical studies, Barrett and his colleagues aimed to explore mesmeric, spiritualistic, and other supernormal phenomena in a rigorous, systematic way. Scientists affiliated with the society were especially interested in accounts of thought transference, because unlike the haunted houses and spirit rappings that many people wanted to investigate, the transmission of images suited the format of psychological experiments. As Barrett had told the Royal Society, he wanted to see "how far recognized physiological laws would account for" unexplained experiences of thought transmission (Gurney, Myers, and Podmore 1: 13).

Because of the long history of fraud in studies of psychical phenomena, members of Barrett's society strove to make their experiments as "scientific" as possible. Early volumes of the *Proceedings of the Society for Psychical Research,* the journal in which they published their investigations, contain rigorous statistical analyses comparing the achievements of mind readers to the results that might be obtained through random guessing. In studies of thought transference, proposed Gurney, "questions of mood, of goodwill, of familiarity, may hold the same place . . . as questions of temperature in a physical laboratory" (Gurney, Myers, and Podmore 1: 30). Determined to avoid duplicity, investigators blindfolded their thought readers and stuffed their ears with putty. They never used paid mediums, observing all experiments themselves, and when necessary they personally supplied the thoughts their readers were to receive. The responsible investigator of psychical phenomena needed to monitor the mental state of all participants, Gurney warned, just as the physicist needed to know the "condition of his instruments." He did acknowledge, however, that the scientist could not "expect in the operations of an obscure vital influence the rigorous certainty of a chemical reaction" (Gurney, Myers, and Podmore 1: 51).

Because of their determination to be scientific, members of the society clashed over Spiritualism, or communication with the dead. Although the Society invited several prominent Spiritualists to serve on its first council, they quickly resigned their positions on seeing the society's contemptuous dismissal of their field (Doyle 2: 64). F. W. H. Myers, however, one of the society's most influential members, believed that if living minds could transmit thoughts among themselves, they should also be able to enjoy "sweet converse with dear departed friends" (Doyle 2: 69). Popular experience with Spiritualism had preceded Barrett's more "scientific" studies of thought exchanges among the living, and despite investigators' efforts to distinguish them, the two types of contact merged in their very definitions. The essential principles of Spiritualism, Doyle wrote, were "continuity of individuality and power of communication" (2: 80). If reports of contact with the dead threatened to undermine more rigorous studies of thought transmission, they also paved the way for them, suggesting a continuous network of connections among minds.

This wireless web was also inspired by technological networks. When Myers coined the term "telepathy" in 1882, he undoubtedly had the telegraph in mind. Thinking analogically, he relied on his society's "writing at a distance" to define the new concept of "feeling at a distance" (Barrett 68). In their second report on thought transference, Gurney, Myers, and Barrett defined it as "a mental perception . . . of a word or object kept vividly before the mind of another person or persons, without any transmission of impression through the recognized channels of sense" (Gurney, Myers, and Barrett 70). Their repeated references to these "channels" suggest a hydraulic, perhaps even a machinelike concept of the nervous system, in which information flows toward a central processing apparatus.[1] Gurney, who wrote most of the committee's reports, thought hard about what to call people's unexplored capacity for communication. The traditional name, "thought-reading," threatened to alarm the public by suggesting that readers would "probe characters and discover secrets" (Gurney, Myers, and Podmore 1: 10). From Gurney's cultural standpoint, it appeared that people could communicate not just because of open channels but because of the barriers between them. "To suppose that people's minds can be thus open to one another," he reflected, "would be to contradict the assumption on which all human intercourse has been carried on" (Gurney, Myers, and Podmore 1: 10). If human minds held nothing back, there would be no individuals among whom to establish connections. Communication required restrictive boundaries that could be breached in carefully controlled ways.

Throughout the eighteenth and nineteenth centuries, the boundaries of self had been penetrated by mesmerists, and the study of thought transmission in the 1880s took as much from these earlier "magnetic" fusions of minds as it did from Spiritualism or telegraphy. The idea that one mind could communicate directly with another grew out of observations made during mesmeric experiments. Some mesmerists claimed that their "vital influence" could "act over great distances," and scientists observing the behavior of people in mesmeric trances found it plausible that "thoughts might be transmitted by the action of a powerful will upon sensitive brains at a distance" (Barrett, Gurney, and Myers 32; Myers, "On Telepathic Hypnotism" 127). Many experimenters reported that the people they hypnotized experienced their sensations exactly as the hypnotists did and reacted to unpleasant tastes or sudden pains as though they were feeling them themselves. In the mesmeric trials that Barrett observed, he noticed a "community of sensation" between the mesmerist and his subject (Barrett 65). During these experiments, the participants' minds merged to the degree that any physical stimulus the mesmerist experienced was communicated to the patient as well. The traditional concept of mental rapport between mesmerists and their subjects, Gurney speculated, might well be rooted in thought transference (Gurney, Myers, and Podmore 1: 11).

In 1842, surgeon James Braid coined the term "hypnotism," indicating a type of mesmerism in which subjects freely suspended their will due to sensory stimulation. He tried hard to distinguish hypnotism from ordinary mesmerism, in which a force projected by the mesmerist was thought to create the trance (Winter, *Mesmerized* 184–85). In the 1880s, hypnotism gained respect as a therapeutic method. In their eagerness to make their own field more reputable, members of the Society for Psychical Research cited new hypnotic studies and tried to collaborate with the psychologists conducting them, many of whom were interested in psychical phenomena themselves. In 1886, Pierre Janet reported that one of his patients could be hypnotized from a distance, and he and Myers systematically studied how her susceptibility to "telepathic hypnotism" varied with spatial separation. To Myers, this prominent scientist's observation of "production of sleep and other hypnotic phenomena by the will . . . of a person at a distance from the subject" provided the strongest possible evidence for telepathy ("On Telepathic Hypnotism" 127).

The possibility of one strong-willed person acting on another had been used for some time to explain certain puzzling human behavior. In the popular "willing game," a subject would leave the room; a group of friends would then think of a task to be performed and "will" it with all of their might. The subject, after reappearing, very often accomplished the desired task, such as locating a hidden object. In such experiments, wrote T. A. McGraw, people were witnessing "the possibility of the nervous system of one individual being used by the active will of another to accomplish certain simple motions" (Barrett, Gurney, and Myers 15–16). When minds merged, it was likely that one would dominate and direct the other, so that the person with the stronger will operated the other via remote control.

The most respected physiologists of the day believed that the willing game involved something very different. In cases where there was no conscious deceit, they proposed, honest players who wanted the game to succeed were guiding the subject with unconscious bodily movements. Often players touched subjects when they reentered the room in order to communicate their collective will. What passed for mind reading, the physiologists claimed, was actually "muscle reading," for the impressions conveyed through this touching were not merely mental. Even when there was no physical touching, players might offer cues through unconscious gestures or movements of their lips. The highly respected physiologist William Carpenter (1813–85) wrote that in the willing game, "communications are made by unconscious muscular action on the part of one person and automatically interpreted by the other" (Barrett, Gurney, and Myers 14). It was for this reason that investigators like Barrett blindfolded their subjects and plugged up their ears. In order to prove the existence of telepathy, they had to rule out the more scientifically acceptable hypothesis of muscle reading.

During the 1880s and 1890s, the Society for Psychical Research gained increasing public attention, and both scientists and nonscientists sometimes found Barrett's interpretations of mental phenomena more appealing than those of mainstream physiologists. Most of the subjects, as well as the investigators observing them, genuinely believed that they were receiving thoughts from other minds. Contrary to what one might expect, explained Gurney, in telepathy, the person being read—the "agent"—was the active party, concentrating ferociously on a single impression to be transmitted. The reader, or "percipient," had only to clear his or her mind and let the foreign impression take shape (Gurney, Myers, and Podmore 1: 10–11). Successful percipients reported that they had to exert considerable effort to achieve the "inward blankness" necessary to receive other people's thoughts. Paraphrasing his subjects' descriptions, Gurney wrote that as one tried to clear one's head, "images [were] . . . apt to importune the mind, and to lead to guessing; the little procession of them marches so readily across the mental stage that it is difficult to drive it off, and wait for a single image to present itself independently" (Gurney, Myers, and Podmore 1: 34–35).[2] To receive impressions transmitted by another mind, one had to be both active and passive, expending energy to silence one's own variable inner voices.

Like physiologists investigating individual nervous systems, investigators of telepathy wanted to know what exactly was being transmitted. Experiments investigating thought transmission—like experiments that measured the velocity of nerve impulses—promised to reveal the nature of thought or at least something about the nature of the information transmitted. In their very first report on thought transference, Barrett, Gurney, and Myers wondered about the relative roles played "by mental *eye* and mental *ear*" (Barrett, Gurney, and Myers 28, original emphasis). "In what form," wondered Gurney, "[is] the impression flashed upon the percipient's mind?" (Gurney,

Myers, and Podmore 1: 28). To answer these questions, investigators had to rely on the percipients' subjective descriptions, none of which was particularly informative. The five clairvoyant Creery girls, who could fetch objects that Barrett or Myers telepathically "ordered," claimed that they suddenly "seem[ed] to see" the things requested (Barrett, Gurney, and Myers 28). "Miss X," who analyzed her unusual mental abilities in a report of 1895, wrote that it was not a question of clairvoyance, coincidences, guesses, or even telepathy. She just suddenly knew something, she explained, and "it came into [her] head to say it" (Miss X 117). If one trusted the percipients, the information received did not correspond to any known sensory modality.

Far more revealing than mind readers' descriptions were their frequent mistakes, which promised to explain the kinds of impressions they were receiving. Percipients were asked to reproduce numbers, words, and sometimes simple drawings, and their partial failures suggested what the agents were transmitting. Often subjects "got" components of words, mistaking Chester for Leicester or Freemore for Frogmore, and Barrett once "successfully obtained a German word of which the percipient [who could not read or write German] could have formed no visual image" (Gurney, Myers, and Podmore 1: 28–29). Such results suggested the transmission of sounds, but near misses in cases involving numbers, drawings, and playing cards implied the communication of visual impressions (Barrett 67). Describing experiments in which he had acted as agent, Barrett told the Royal Society in 1876 that "the existence of a distinct idea in my own mind gave rise to an image of the idea in the subject's mind; not always a clear image, but one that could not fail to be recognized as a more or less distorted reflection of my own thought" (Barrett 74). Convinced that minds communicating telepathically were transmitting visual impressions, Barrett claimed that mind reading offered "a far more perfect interchange of thought than . . . the clumsy mechanism of speech" (Barrett 69). In efficiency, telepathy surpassed telegraphy, which had not yet transcended the primitive, organic vehicle of spoken language.

Understanding the mode in which thoughts were received, of course, revealed nothing about their form while traveling *between* the agent's and percipient's heads. By what physical mechanism were these impressions being transmitted? In their attempts to answer this question, investigators of psychic phenomena relied heavily upon telegraphy. When they invoked models offered by contemporary physics, they raised the same issues as scientists studying the nerves or designing technological communications systems. Some psychical investigators believed that the information traveled as waves, others as particles.

In their initial report on thought reading in 1882, Barrett, Gurney, and Myers proposed that "for every thought there is a corresponding motion of the particles of the brain, and . . . this vibration of molecules of brain-stuff may be communicated to an intervening medium, and so pass under certain circumstances from one brain to another, with a corresponding simultaneity of impressions" (Barrett, Gurney, and Myers 33). Reasoning from analogy in a letter to these investigators, Charles Ede asked, "may not a material vibration in a strong brain affect another by its vibration, as light at a distance acts upon the retina of the eye, or sound upon the ear?" (Barrett, Gurney, and Myers 32). While this characterization of light as a vibration of particles suggests eighteenth- and not nineteenth-century physics, the concept of an intervening medium, the ether, was accepted by most physicists of the 1880s. Many scientists, including the influential biologist Ernst Haeckel (1834–1919), believed that thought consisted of altered vibrational patterns.

Just as Emil DuBois-Reymond had used Michael Faraday's models to explain nerve-impulse conduction, some scientists of the 1880s proposed that thoughts might be transferred via "nervous induction" (Gurney, Myers, and Podmore 1: 16). T. A. McGraw, who in 1875 had proposed that one person could "use" the nervous system of another, offered a mechanism through which this exploitation might occur:

> [There are] strong similarities that exist between nervous and electrical forces, and, as we know, it is possible to generate induced currents of electricity in coils of wire that are near to a primary electric coil; so we can imagine the nervous current to be continued into (induced in?) another body and act there upon the automatic centers of action. (Barrett, Gurney, and Myers 16)

Because nervous impulses consisted of electrical signals and obeyed physical laws, it was reasonable to expect that they could be impressed upon other minds. Such an effect would of course decline with distance, but if one identified thought with electricity, "nervous induction" seemed perfectly reasonable.

Technological progress offered the best argument for telepathy. Between 1894 and 1896, Italian physicist Guglielmo Marconi (1874–1937) developed the wireless telegraph, proving that people could transmit information through the atmosphere without relying on a physically continuous conductor. Because Marconi collaborated with British physicists and patented his invention in England, the British public quickly learned about the "wireless" and began speculating about its implications.[3] To investigators of psychic phenomena, it made thought transference more plausible than ever.

William Barrett asserted that:

> the existence of wireless telegraphy and the bridging of vast spaces by mes-
> sages transmitted in this way naturally suggest that thought might likewise be
> transmitted by a similar system of ether waves, which some have called
> "brain waves." And there is no doubt the fact of wireless telegraphy has
> made telepathy more widely credible and popular. (Barrett 107)

While wireless telegraphy provided the best analog for thought transference,
it was the original, wire-bound system that gave telepathy its name and general concept. Whether late nineteenth-century investigators of thought
transmission explained it with particles or waves, their descriptions of
telepathy always involved the instantaneous communication of impressions
through energy fluctuations at the molecular level.

But even as they offered physical mechanisms for telepathy, members of
the Society for Psychical Research spoke out against materialistic explanations. They argued for thought transference from two directions at once,
offering plausible physical models they hoped would appeal to scientists
while asserting that psychical phenomena could never be explained by physical laws. William Barrett, for instance, claimed that "telepathy renders a
purely materialistic philosophy untenable" (Barrett 69).

After comparing thought transmissions to the transmissions of wireless
telegraphs in 1911, Barrett attacked his own analogy. As a form of energy,
he argued, brain waves should follow an inverse square law, rapidly dimin-
ishing with separation from their source. Consequently, it would require
"tremendous energy" to transmit them over significant distances. Studies of
telepathy, however, showed no regular relationship between successful
transference and spatial separation, and agents transmitting over great distances never experienced a great expenditure of mental energy. For Barrett,
these facts suggested that "supernormal phenomena . . . do not belong to the
material plane, and therefore the laws of the physical universe are inapplic-
able to them" (Barrett 108–9). Despite his tremendous efforts to found a
"scientific" field of study, the investigator who had brought thought transference before the Royal Society ultimately rejected physical explanations,
characterizing mental telepathy as the direct communication of impressions
from soul to soul.

According to Barrett's findings, these transmissions were identical to
messages circulating *within* an individual soul. When Barrett and his colleagues described the workings of the mind, they never presented individual
consciousness as a simple, unified entity. F. W. H. Myers, with his deep
interest in subconscious phenomena, believed that in daily life, people con-

stantly received "messages communicated from one stratum to another stratum of the same personality." The "monitory voices" that some people reported hearing were actually "messages from the subjacent self" (Myers, "Subliminal Consciousness" 298). Within the individual mind there existed barriers penetrated by channels of communication, so that people who learned to read others' minds were really building upon experiences they had acquired through reading their own.[4] Like Myers, Barrett believed that the unconscious—not the conscious—mind was responsible for thought transmissions. Because minds were so divided, people received and deciphered strange impressions throughout their daily lives. If they learned how to control their internal communications, they would eventually learn how to communicate with other minds. "If we knew how to effect this transfer," reflected Barrett, "unfailingly and accurately, from the outer to the inner self and *vice versa,* telepathy would become a universal and common method of communicating thought" (Barrett 110). The secret to developing connections between minds was realizing that minds were open and dynamic, a collection of complex networks waiting to be interlinked.

Mental Telegraphy: The Telepathic Fictions of Mark Twain

Mark Twain (1835–1910) took great pride in being one of the first private individuals ever to own a telephone. He was also one of the first well-known authors to use a typewriter (Ketterer xxii). A lover of new inventions, Twain made a number of his own, including a "perpetual calendar," "a board game for teaching history," and a "bed clamp" to prevent small children from kicking off their blankets. Intrigued by machines, he was a "patent office habitué," and he maintained a special interest in communications technology (Ketterer xxii).

Twain's early work as a printer, reporter, and long-distance correspondent gave him a lifelong interest in disseminating the written word. When his father died, he went to work as a printer's apprentice at the age of twelve. In school he excelled at nothing but spelling, but when he began typesetting, he had extensive opportunities to put this skill to work. His supervisor reported that "in setting type and printing Sam was both swift and clean" (Gerber 4).[5] Twain never lost interest in the mechanics of printing, and later in life he went bankrupt after investing his entire fortune in a new technology for typesetting (Ketterer xxiii). This attempt to improve dissemination of the written word would also hurt another investor—Twain's friend Bram Stoker.

As a journalist during the years between 1847 and 1869, Twain welcomed any technology that helped spread information more rapidly. He had great interest in the telegraph, which was just then reaching the Midwestern and Western cities where he wrote. In his profession, the ability to transmit information was all-important, and Twain sometimes had to resort to creative means. While in Hawaii, he once got a scoop on a story about shipwreck survivors by "having [his manuscript] thrown onto the deck" of a boat about to leave for the mainland (Gerber 20). It would certainly have been more convenient to transmit his thoughts directly to the San Francisco paper or, for that matter, to the paper's entire circulation. As a long-distance correspondent, Twain was doing exactly this, offering eyewitness accounts of distant events so that readers could reconstruct them in their imaginations. Given the disheartening delays in disseminating these images, it is hardly surprising that Twain dreamed up more efficient ways of transmitting them.

While Twain maintained his interest in communications technology, he was forever fascinated by claims about what the mind could do. Intrigued throughout his life by the possibility of thought transmission, he once tried to contact his dead brother Henry and daughter Susy by using a spiritual medium (Ketterer xxvii). At the same time, Twain distrusted all claims of supernormal powers, having spent a lifetime studying human gullibility. In his story "Mental Telegraphy" (1891), Twain's narrator avers, "I have never seen any mesmeric or clairvoyant performances or spiritual manifestations which were in the least degree convincing" (102). Twain himself once "faked hypnosis so dramatically" at a local demonstration that the lecturer asked him to come back the next night (Gerber 3).

"Mental Telegraphy" reflects these conflicting feelings of skepticism and enthusiasm. At times Twain's tale reads much like reports to the Society for Psychical Research, arguing earnestly from analogy and personal experience. As Twain himself did, the narrator claims to have written the bulk of his story in 1878, then to have "pigeonholed" it for years, fearing that the public would not take it seriously. He explains that he has decided to publish it in 1891—again, as Twain actually did—because of "the flood of light recently cast upon mental telegraphy by the intelligent labors of the Psychical Society" (97–98). It is never clear, though, to what extent Twain can be identified with this narrator. Twain developed as a writer by assuming personae, and many of his early journalistic efforts consisted of humorous letters to the editor from fictional correspondents (Gerber 10–15). While "Mental Telegraphy" raises powerful questions about the direct transmission of thoughts, it is told tongue in cheek, and it laughs at the phenomena its narrator fears will be ridiculed by the public.

Initially, Twain presents mental telepathy as a plausible explanation for events dismissed as coincidences. He offers a great variety of examples, both intriguing and funny. Beginning with the idea of letters crossing, the narrator asserts that such "curious coincidences" are "no more accidental than is the sending and receiving of a telegram an accident" (96). Often, he claims, he has suddenly known that he would receive a letter from someone from whom he has not heard in years, and he describes several instances in which he disclosed the contents of such letters to astonished friends without ever having opened them. As soon as we mention long-absent people, he writes, they frequently appear, and he claims to have become more circumspect after criticizing people, only to find that he was talking with their relatives (99). The déjà vu phenomenon, he avows, is not evidence for a previous existence, but for "the fact that some far-off stranger has been telegraphing his thoughts and sensations into your consciousness, and that he stopped because some countercurrent or other obstruction intruded and broke the line of communication" (103). If one accepts the fact that minds can "wire" each other, he submits, all of these puzzling phenomena have a very reasonable explanation.

As further examples of mental telegraphy, the narrator offers rival claims to scientific inventions, many of which are related to communications technology. Foremost among them are Joseph Henry's, Samuel Morse's, Charles Wheatstone's, and Karl August Steinheil's designs for a telegraph in 1837 (101). Twain's list of concurrent discoveries in science includes theories of natural selection, differential calculus, light waves, spectral analysis, and the interpretation of Egyptian hieroglyphics. Interestingly, almost all of these examples involve the recording or movement of information, so that the content of the messages "cabled" from one scientist's mind to another reflects Twain's hypothesis about how the ideas are conveyed. "Perhaps," he proposes, "*one* man in each case did the telegraphing to the others" (107). In the case of Thomas Young and Jean François Champollion struggling to decipher hieroglyphics, the mental telegrams would have encoded messages about codes.

Proposing that "mind can act upon mind in a quite detailed and elaborate way over vast stretches of land and water," Twain's story sometimes appears to offer a scientific argument (96). Like T. A. McGraw, the narrator submits that the something which conveys our thoughts through the air from brain to brain is a finer and subtler form of electricity, and all we need do is to find out how to capture it and how to force it to do its work, as we have had to do in the case of the electric currents. Before the day of telegraphs neither one of these marvels would have seemed any easier to achieve than the other. (106)

If eighteenth-century scientists learned to control electricity, the narrator implies, why should nineteenth-century scientists not learn how to harness analogous forms of energy? Like a telegraph, he proposes, a mind can send and receive thoughts, and like a telegrapher, it translates between written language and its own signals.

Echoing the desires of William Barrett, Twain's narrator urges that this new type of telegraphy progress further, transcending the language in which communication is conventionally conducted. Suppose, writes Twain's narrator, "that mind [can] communicate accurately with mind without the aid of the clumsy vehicle of speech" (105).[6] Because telegraphy relies on spoken words, its rate-limiting steps are those of encoding and decoding, whereas the transmission itself is instantaneous. Even the direct transmission of speech strikes the narrator as inadequate. "The telegraph and the telephone are going to become too slow and wordy for our needs," he predicts. "We must have the *thought* itself shot into our minds from a distance; then, if we need to put it into words, we can do that tedious work at our leisure" (105–6, original emphasis). An expert manipulator of words, Twain laughs at mental telegraphy, at readers swayed by his arguments, and at the "clumsy vehicle" of language by which he and his readers are forced to communicate.

Perhaps the most fascinating aspect of Twain's story is the relationship it proposes between language and thought. Taking the telegraph as a model for the mind, the narrator presents thought as one thing, language as another. Explaining how he knew beyond all doubt what was in a friend's letter, he tells us: "I think the clairvoyant proposes to actually *see* concealed writing. . . . I only seemed to know, and to know absolutely, the contents of the letter in detail . . . but I had to word them myself. I translated them, so to speak" (100, original emphasis). Like percipients describing their experiences to the Society for Psychical Research, Twain's narrator can express the ideas he has received only after converting them into language. As a sensitive, perceptive apparatus, he implies, the human nervous system intercepts and translates an endless succession of signals. One morning, he reports, "ideas came pouring into my head from across the continent, and I got up and put them on paper" (102).

While the mind resembles a mechanical receiver, Twain's narrator suggests that it transmits and transduces much more efficiently than the telegraph. In the frequent cases in which one suddenly knows something before the telegram arrives to announce it, he claims that the "telegram has gone straight from your brain to the man it was meant for, far outstripping the wire's slow electricity" (108). In Twain's depiction of mental telegraphy, the body's communications system regains its superiority over the aging technologies modeled upon it. The new "phrenophone" he proposes for transmitting thoughts emulates the mind's telepathic abilities just as the telegraph emulated the nerves, the organic once again suggesting a design for the mechanical transmitter.

But the very concept of mental telegraphy was inspired by evolving technology. In a later story, "From the *London Times* of 1904" (1898), Twain develops his anecdotes about mental telegraphy into a tall tale about the clash of science and politics. Like "Mental Telegraphy," this story is recounted by a fictional writer, a correspondent to the *Times* whose observations one cannot fully believe. Szczepanik, an eccentric inventor, has produced a "telelectroscope" that offers simultaneous voice and visual contact with any region in the world. Unfortunately, his financial backer, Clayton, does not share his high opinion of the device and bets a friend that it will never be worth a brass farthing to anyone. The inventor and backer fight, and later, when Szczepanik is killed, Clayton is blamed for his murder and imprisoned.

Alone in his cell, Clayton begins using the telelectroscope and revels in the same technologically aided mental travel described by telegraph operators decades earlier. With his space and time strictly delimited, the condemned man begins surfing the net. "He called up one corner of the globe after another," describes the narrator, "and looked upon its life, and studied its strange sights, and spoke with its people, and realized that by grace of this marvelous instrument he was almost as free as the birds of the air, although a prisoner under locks and bars" (130). Like Thayer's and James's telegraphers, he is tantalized by virtual freedom while locked in a cage.

Spatially confined, his hours dwindling, Clayton dials into a coronation in China, where a speaker explains to him: "This *is* yesterday—to you" (131). He regrets that he must miss the end of the ceremony because of a "previous engagement," but the narrator—the supposed eyewitness correspondent to the *London Times,* who is with him in his cell—continues to watch. Just as Clayton is about to be hanged, the narrator screams for the executioners to halt, for he has spotted Szczepanik in the Chinese crowd.

Tired of notoriety, the inventor has faked his own death. Clayton is found innocent, and thanks to the technology he scorned, he is saved.

Unfortunately, human stupidity triumphs over communications technology. According to a French precedent, a man cannot be pardoned for a crime he has not committed, so the innocent Clayton is hanged anyway. The telelectroscope offers dazzling new possibilities for dialogue and cultural exchange, but irrationality and violence continue to dominate human relations.

In each story, Twain inspires readers to think about where telecommunications are heading, and he draws an intriguing parallel between the body's and the telegraph's transmissions of thoughts. So ironic is his tone, however, that one can hardly believe he has received or transmitted any mental telegrams. Early in "Mental Telegraphy," the narrator promises to relate "an extraordinary experience of mine in the mental telegraphic line," preparing the reader for a tall tale (96). Twain's examples themselves provoke laughter, particularly the ones about scientific and literary priority. When he discusses disputed claims to original ideas, his tone becomes deeply ironic. Mentioning several astronomers' simultaneous discoveries of an unknown planet, Twain asks, "if one astronomer had invented that odd and happy project fifty years before, don't you think he would have telegraphed it to several others without knowing it?" (107). While his narrator argues that "coincidence" is a "cheap and convenient way of disposing of a grave and very puzzling mystery," Twain suggests through his humorous tone that mental telegraphy is the convenient label, masking either plagiarism or a widespread interest in scientific questions (99). "Consider for a moment how many a splendid 'original' idea has been unconsciously stolen from a man three thousand miles away!" he exclaims. "Is it not possible that inventors are constantly and unwittingly stealing each other's ideas whilst they stand thousands of miles asunder?" (101). Twain's implications become clearest when he recalls "lively newspaper wars" in which three or four authors all claimed to have written the same poem. "These were all blameless cases of unwitting and unintentional mental telegraphy, I judge," comments his narrator innocently (102).

At moments like this one, Twain mocks investigators of telepathy who promised "scientific" explanations for events dismissed as coincidences. But the very fact that Twain wrote the stories, taking pains to parody writers like Barrett and Gurney, indicates his—and his readers'—fascination with unexplained thought transmissions. Daily use of the telegraph and growing frustration with its limitations encouraged people to reexamine the body's ability to transmit information, looking for hints about how communications technology might be improved. At the same time, technology suggested the mind's hidden potential to communicate. Together, the organic and the technological systems pointed toward a wireless world.

Bram Stoker's Communications Net

Bram Stoker (1847–1912) knew Mark Twain quite well. In the early 1890s, their friendship foundered when Twain persuaded Stoker to invest money in a new technology for typesetting. In letters to Stoker, Twain boasted of wonderful new typesetting machines, and Stoker bought a hundred dollars worth of stock. When the venture failed, Twain returned Stoker's money, but they were never again on good terms (Farson 74–75).

If one considers Stoker's development as a writer, it is easy to see why typesetting appealed to him as an investment. Preparing for a career in the Irish Civil Service, he studied science at Trinity College and was graduated with honors in 1870. Stoker began working as a legal clerk in Dublin Castle, but his ambition never matched his passion for theater. In late 1871, he began writing as a drama critic for the *Dublin Mail,* and his intense friendship with charismatic actor Henry Irving led to Irving's request that he manage the Lyceum Theater (Farson 24–25). In the fall of 1878, Stoker abandoned his career in the Irish Civil Service to begin supervising Irving's elaborate productions.

As the man responsible for the Lyceum, Stoker became the central nexus in a communications web through which vast quantities of information flowed.[7] He managed finances, organized tours, and exchanged thousands of letters and telegrams with people throughout Europe and the United States. In an ongoing struggle against disaster, he fought to bring productions together. Often, Stoker wrote as many as fifty letters in a day, and there were times when he "lived in trains" (Farson 43, 181). Relying utterly on the post, the railway, and the telegraph, he specialized in acting at a distance.

Because of Stoker's work, telegrams became a central part of his existence. Irving frequently summoned him by telegram, and it was through such summons that he learned of the job offer as theater manager and of Irving's knighthood in 1895 (Farson 38, 175). Stoker's activities kept him apart from his young wife and child a good deal, so the telegraph played an important role in his personal life as well. Once, when his wife and son were shipwrecked off the French coast, he believed for two days that they had drowned because the telegram announcing their rescue was mislaid (Farson

103). When the company gave a command performance at Windsor, the royal family placed the castle's telegraph office at Stoker's disposal, and his dispatches were paid for by the queen (Farson 93). As manager, Stoker was bombarded with information, and it was his job to deal with the hundreds of messages that poured into his theater each week.

If the Lyceum's twenty successful years are any indication, Stoker performed his tasks well, and he always kept himself informed about day-to-day matters. According to his workers, he "controll[ed] his staff of 128 like a regiment" and had the ability to "see through a brick wall" (Farson 57). Particularly telling is Stoker's bookkeeping method: he relied on "reticence," always "playing one department against the other, so that each might have a glimmering in part but none knew the whole" (Farson 44). Stoker and his theater company survived because of his ability to transmit and retain information. He succeeded not just because of the messages he sent but because of what he chose not to pass on, and he incorporated this survival technique into his most successful novel.

Written while Stoker was managing the Lyceum, the text of *Dracula* reproduces the communications net in which he lived his daily life.[8] But as an "organized" text, a system of interconnected parts, the novel also pulses like a body. Like Twain, Stoker incorporates the format of long-distance communications into his story's structure. Writing of circulation and transmission, he tells the tale through letters, transcribed journals, newspaper clippings, and telegrams. Just as Twain does in "Mental Telegraphy," Stoker opens his story with an explanatory note from a fictitious person. An unknown editor claims that in the text one is about to read, "all needless matters have been eliminated" and that "all the records chosen are exactly contemporary" (5). This, of course, arouses the reader's suspicions about who chose these particular documents and why. The editor's opening promise—"how these papers have been placed in sequence will be made manifest in the reading of them"—introduces one of *Dracula*'s central issues (5). Although the novel describes a vampire hunt, it is also a story about how to construct a narrative.[9]

Apart from Dracula himself, the greatest problem Stoker's characters face is how to put their documents in order. They do not attempt the hunt until they have finished collating, organizing their various accounts of the vampire into a single, typewritten narrative (Moretti 438). Even Stoker's earliest notes for the novel, from August 1890, indicate that he had always planned to open it with a legal correspondence (Stoker, "Notes for Dracula" 340). Stoker describes occult phenomena with the imperfect communications techniques of everyday life, and his characters are as preoccupied as he is about how to represent the unspeakable. After the solicitor Jonathan Harker and his helpful wife, Mina, have spent an entire day collating, they reassure themselves that "now, up to this very hour, all the records we have are complete and in order" (208). The characters' ordering of their documents reduces their fear of forces they cannot control.

But what is order? How does one "organize" papers; how does one tell a story? Confronted with the facts by "open-minded" psychologist Van Helsing, his younger colleague Seward despairs: "I am going in my mind from point to point as a mad man, and not a sane one, follows an idea. I feel like a novice blundering through a bog in the mist, jumping from one tussock to another in the mere blind effort to move on without knowing where I am going" (172). One can connect recorded events through any number of leaps, Stoker implies, but as Eliot's *Middlemarch* suggests, not all connections are equally valuable.

Mina Harker, who, like Stoker, coordinates the characters' communications, tells Seward that "in this matter dates are everything" (198). As the editor within the text, she suggests putting the hunters' accounts into chronological order, and she types in triplicate what will become the typescript of the novel. By using time as the organizing parameter, Mina—and Stoker—are "able to show a whole connected narrative" (Stoker, *Dracula* 199). Later in the novel, Jonathan Harker provides an internal critique of their efforts, declaring that "the whole story is put together in such a way that every point tells" (218). While Mina has the most trustworthy scientific—and literary—instincts, it is the psychologist Seward who reflects the most about what they have accomplished in compiling this case study. Once Mina imposes chronological order, the events become a timetable of sorts. Seward sees that "all [his patient Renfield's] outbreaks were in some way linked with the proximity of the Count" (200). When they are arranged chronologically, the characters' subjective descriptions become both literature and science, for they reveal the connections of cause and effect.[10]

Pleased with their *post hoc ergo propter hoc* logic, the vampire hunters take as much pride in their record keeping as in the chase itself. Through much of the novel, their writing is far more successful than their hunt. As the men pursue Dracula, the production of the collated text is left increasingly to Mina, who seeks "some thread of continuity" in the disparate accounts (161). Mina keeps material to be typed in her workbasket, so that her collation replaces the traditional female task of sewing. As the hunters' collective wife and computer, she "knit[s] together in chronological order every scrap of evidence they have" (199).[11] In some sense, Stoker's text is as much a web as Eliot's, a patchwork of interwoven scraps. Like Eliot, Stoker and his characters strive to create a reasonably faithful history by connecting the observations of many different individuals, and they create a narrative about their own weaving. Like *Middlemarch, Dracula* is a novel about communicating and, more particularly, about writing.

Writing about Communicating, Communicating about Writing

Except for Dracula himself, who is given no narrative voice, all of Stoker's characters think consciously about their own writing. Their transcriptions allow them to maintain their sanity and distribute information, and their typing and collating become the activity of a collective brain. Yet even while given little opportunity to express himself, Dracula writes as well. Whereas the hunters write on paper, he writes on bodies. Mina's scar and Lucy's neck wounds are the vampire's own marks, the records of his strategies for control.[12] As the hunt proceeds, it becomes clear that the Count's and his opponents' communications nets are mirror images of each other. As the spider in a rival communications web, Dracula performs all of the same activities as the hunters, relying on corporeal rather than technological connections. While the Count's network is primarily organic and the British hunters' is technologically based, the two systems correspond in a great many ways. Both carry out the activities of a vast nervous system, perceiving, encoding, transmitting, recording, and recalling information.

Several of the vampire hunters find writing soothing. Trapped in Dracula's castle, Jonathan Harker confesses, "feeling as though my own brain were unhinged or as if the shock had come which must end in its undoing, I turn to my diary for repose. The habit of entering accurately must help to soothe me" (41). For Harker, writing provides a way to maintain control over experiences that threaten his worldview.[13] Mina, too, finds that writing calms her, confessing that "it soothes me to express myself here; it is like whispering and listening at the same time" (72). By providing psychological feedback, personal writing offers an illusory support network, reassuring one—falsely—that one is not alone in one's thoughts.

Both Harker and the senior psychologist Van Helsing are most inclined to write when they doubt their own sanity. Helpless with guilt and rage when he learns that Dracula has mingled with Mina, Jonathan declares, "as I must do something or go mad, I write this diary" (252). Van Helsing confesses to an unknown reader, "You may at the first think that I, Van Helsing, am mad—that the many horrors and the so long strain on nerves has at the last turn my brain" (315). Stoker's characters believe they are writing to save their lives—or at least, their minds—and their writing becomes their strongest defense against Dracula's disorder.

Writing would thus appear to be associated with sanity, but in *Dracula,* the mad write, too. Seward notes that Renfield "keeps a little notebook in which he is always jotting down something," and the lunatic maintains careful records of his zoophagous feats (69). As Stoker describes it, Renfield's writing becomes a wonderfully ironic reflection of the psychologist's own note taking. Seward envisions a God who is forever watching him, a "Great Recorder [who] sums me up and closes my ledger account with a balance to profit or loss" (71). Stoker's madman and psychologist share this divine role, each creating his own account of the other.

When Mina begins her journal, she too appears to be seeking personal satisfaction. She tells her friend Lucy that she is writing for herself, not for others, but she quickly finds herself "mak[ing] my diary a duty" (86). Eventually, Mina becomes the hunters' "scribe" and "Recording Angel," turning everyone's experiences into coded information that can be quickly transmitted and read (297, 284). The other characters, in contrast, act more like sensory nerves than central nuclei. Harker declares, "I must keep writing at every chance, for I dare not stop to think. All, big and little, must go down; perhaps at the end the little things may teach us most" (253). While their writing and interpreting suggest the activity of a sophisticated nervous system, no character except for Mina sees the full significance of the information at the time it is being recorded. In this nervous system, unlike the one that Henry James depicts, the neurons with direct access to the world know the least. Only the integrative center sees all and draws conclusions accordingly.

The characters love their own writing because of the power it gives them to record and transmit information, but they also love it for what it conceals. Unlike the others, Seward records his observations as impressions left by his own voice, reading his case notes onto phonograph cylinders. When Mina jacks into them, "putting the forked metal to her ears," she tells Seward, "That is a wonderful machine, but it is cruelly true. It told me, in its very tones, the anguish of your heart. . . . No one must hear them spoken ever again! . . . I have copied out the words on my typewriter, and none other need now hear your heart beat, as I did" (197). Mina sees words as separable from a speaker's emotions, and by converting vocal impressions into written signs, she believes she can make valuable information public while keeping Seward's feelings private (Wicke, "Vampiric Typewriting"

471). Writing, especially typewriting, involves a splitting and a repression, always cordoning off a part of the experience it claims to encode.

Throughout the novel, Stoker plays with the differences between written and spoken language. In a wonderfully ironic telegram, Arthur Holmwood promises the hunters, "I bear messages which will make both your ears tingle" (62). Even Dracula himself is deeply conscious of the difference between writing and talking, telling Harker, "I know the grammar and the words [of English], but yet I know not how to speak them" (26). Ultimately, he hopes to master both written and spoken English. Because success depends on communicating orders throughout a complex network, the vampire and his hunters know that whoever can communicate in the greatest variety of ways will conquer the field.

Consequently, in their war of communications, the forces of good and evil rely on many different languages and codes. Dracula has mastered numerous languages, and Van Helsing, who mirrors the vampire in several ways, "knows something of a great many languages" as well (18, 308). Stoker lets us know little of Dracula's codes, for not only is he denied a narrative voice but he communicates with human and animal minds directly. The hunters, however, delight in their language, proud of the swiftness with which they can transcribe information.[14]

We learn immediately that Harker's journal is in shorthand, a new technique of transcription that allowed writing to match the real-time flow of speech. Mina learns shorthand so that she can be useful to her solicitor husband, anticipating the letters that she will someday "take." When she hears of Seward's phonographic diary, she cries, "Why, this beats even shorthand!" (195). Mina's delight with the phonographic cylinders offers some insight into the unspoken goals of transcription techniques. The object is to record information as rapidly as it can be received, then to reproduce it, on demand, as quickly and accurately as possible. The coded form should permit efficient dissemination and should be easily translatable into ordinary written and spoken language. When Mina reflects that "there is . . . something about the shorthand symbols that makes it different from writing," she realizes vaguely that shorthand's representational style brings it closer to fulfilling these goals (72). In its ability to record and reproduce speech rapidly, the phonograph comes closer still.

When assessed by these criteria, traditional writing supersedes speech, because it can be stored, distributed, and reproduced more exactly—as long as there is paper and a supply of competent scribes. Writing falls short of speech, however, in its slowness of transcription and reproduction. Shorthand compensates for this disadvantage, packing more information into every stroke. Both shorthand and writing require special skills, and not everyone can use them to record and transmit thoughts. As in Helmholtz's conception of the nervous system, signals become meaningful not because of their actual form but because of the special properties of the central organ reading them. Harker's shorthand journal proves valuable not because of the speed with which it records impressions but because Dracula can make no sense of it—although he does have the sense to steal all of Harker's remaining paper.

In the hunters' nervous tissue of paper and wire, writing performs a mnemonic function. The characters' written records are their memory and knowledge of the Count, the counterpart to his more intimate knowledge of them. Twice Mina writes that she would like to do "what I see lady journalists do: interviewing and writing descriptions and trying to remember conversations." She has heard that "memory [is] everything in such work— that you must be able to put down exactly almost every word spoken" (56, 163). This, of course, is exactly what she does for the hunters. When Seward lays his hand on the "type-written matter" and declares, "the answer is here," he identifies her typescript as their collective memory (224). Their papers are their brain tissue, and the hunters rely upon them absolutely.

While this paper memory may seem less vital than the body's, Harker's brain fever suggests that both means of recording are vulnerable. Written records can be burned, but neuronal ones can be physiologically disrupted (Wicke, "Vampiric Typewriting" 471). Seeing her husband's fragile mental state, Mina wants him to read Van Helsing's letter, believing that "it may be even a consolation and a help to him . . . to know for certain that his eyes and ears and brain did not deceive him" (162–63). Sure enough, reading an independent account of what he hazily remembers "ma[kes] a new man" of Harker (168). Writing reinforces sanity not just when it is recorded but when it is shared and read, and technologically created memories can help restore those created by a body.

Unlike Dracula, the hunters cannot transmit their thoughts directly, so they must record and copy their written memories as quickly as they can. Here Mina's cutting-edge skill once again comes in handy. As a collective wife, Mina uses her typewriter to reproduce the hunters' memories.[15] Mina has learned typing in conjunction with stenography, hoping to help her husband by converting his speech into a form that can be circulated and read. Like shorthand, typing improves on conventional writing: it is faster, it is legible to all and consequently less noisy, and it allows one to make several identical copies simultaneously. With her dexterous fingers, Mina turns the hunters' letters and journals—and most importantly, Seward's recorded voice—into a "mass of type-writing" (326). She sees her work as a "solemn duty," transforming the hunters' observations so that they are "ready for other eyes" (161). With a typewriter, one can mass-produce individual perceptions so that they quickly reach the eyes of countless others, rapidly communicating impressions from mind to mind.[16]

Memories are worthless if inaccessible, and the greatest advantage Mina's typing confers is accessibility. When she admires Seward's phonographic cylinders, he tells her rather sheepishly, "you see, I do not know how to pick out any particular part of the diary" (196). It is Mina's idea to transcribe his words mechanically, but Stoker uses a corporeal image—always involving the same root word—to convey the "handiness" of this new medium. "I handed [Van Helsing] the shorthand diary," writes Mina proudly. "I took the typewritten copy from my workbasket and handed it to him" (164). At last, she tells us, "I threw myself on my knees and held up my hands to him" (165). While her gesture is one of supplication, the sequence suggests a further symbolic offering. The typist Mina ends by handing over her own hands, the most valuable and versatile instrument in the group's communications technology.

Communicating in the West

The hunters' intelligence network is formidable, not just in its extent but in its diversity. While pointing out their shortcomings and vulnerabilities, Stoker's novel celebrates all forms of modern communication. In *Dracula*, one encounters transmissions conducted by railway, telegraph, mail, newspapers, even a note stuffed into a bottle and a paper concealed in a woman's breast. The Victorian communications network eventually overcomes the invading parasite, but there is a story to be told because its weaknesses keep it from ensnaring him right away.

The British railway system, the great pride of Victorian England, figures prominently in Stoker's novel, and like the author himself, the characters often seem to "live in trains." Like many Victorians, Stoker's characters associate trains with civilization, and when Jonathan dreams of escaping from Dracula, he sees himself running "away to the quickest and nearest train!" (55). A simile of Van Helsing's is even more revealing: "we shall have an open mind, and not let a little bit of truth check the rush of a big truth, like a small rock does a railway truck" (172). In his comparison, the movement of a train becomes "the rush of big truth," so that the railway represents the force of truth itself. The doctor's simile is ironic and complex, for the "big truth" is the existence of Dracula, and the "small rock" is the Western science that denies his existence. In the novel, Western fact and Eastern fact intermingle as the train of Eastern truth rolls over Western tracks.

Dracula opens with Harker's observation that his Austrian train is late, a sign that "we were leaving the West and entering the East" (9). Soon afterward, he comments that "the further East you go the more unpunctual are the trains" (11).[17] Such a violation of the timetable disgusts the young solicitor's clerk, for whom scheduled trips are a fact of daily life. In the nineteenth century, Victorian railway timetables imposed order on time itself, turning hours and minutes into figures on a page.[18] Mina, the "train fiend," memorizes timetables to help her husband, "ma[king] up" first the Exeter-London schedule, then those of Eastern European lines (169, 293). Her knowledge that the next train from Varna to Galatz leaves at 6:30 A.M. does not make a difference in the plot, but she saves the hunters an energy-draining quest for information. By storing facts in her own mental circuits, she keeps her companions fresh for the chase.

The Victorian telegraph network, another source of pride, plays a much more important role in the plot. Indeed, it is questionable whether the hunters could have stopped Dracula without it. As Stoker did daily, the hunters use the telegraph to act at a distance: they order garlic flowers and reserve rooms in Galatz. In the first half of the novel, the telegraph affects the narrative more than the plot, for the characters are merely learning of Dracula's doings after the fact. Collating the text, Mina pastes their messages directly into the typescript, for she regards them as essential elements of their story. At this stage, their telegrams shape the narrative because the flow of the story depends not just on what they know but on when and how they learn it. When Van Helsing first examines Lucy, he tells Seward, "you must send me the telegram every day" (108). Mina learns of Lucy's death through a telegram from Van Helsing; her first acquaintance with him is this terrible dispatch.

Once their telegrams allow them to pool their information, the characters can anticipate Dracula's actions, and the telegraph begins to affect the novel's events. Mina cables Van Helsing to tell him which train to take, and Holmwood telegraphs to ask the whereabouts of the Count's ship. Most significantly, when the men have broken into Dracula's London house, Mina informs them by wire that he left Carfax at 12:45, heading south. Because her information moves more quickly than he does, he fails to take them by surprise. Their clumsy stabs at him accomplish little, but it is a victory for telegraphic over telepathic transmissions.

In tracking the vampire, Stoker's characters also rely on another modern technique for spreading information, the newspaper. Like Twain, Stoker imitates the format of eyewitness accounts, and he incorporates entire fictitious articles into his narrative. Three of these play a significant part in the text: a description of a storm in Whitby; a tale of a wolf who escaped from a London zoo; and an account of a dreadful "bloofer lady," who is attacking children outside of London. Like the characters' journals, all of these articles describe Dracula's deeds from particular perspectives. The first two resemble Twain's fictitious *London Times* story, narrated by eyewitness correspondents who do their utmost to make the reader feel what they are feeling. Trusting the validity of these accounts, Mina has spliced them directly into the text. They are already typescript, already mass-produced information. Even Van Helsing retains his faith in the Victorian media, informing himself of Dracula's doings with a "bundle of papers" he reads on the train (169). While the accounts are subjective, Stoker's characters accept them as official information and weave them into their record unedited. They regard them as extensions of their own perceptions of Dracula, information gathered through adjacent links in their web of intelligence.

While the modern media prove valuable, it is the professional networks that enable both Dracula and his hunters to act at a distance. Harker comes to Dracula, initiating the narrative, because the Count is seeking British agents to facilitate his invasion of England. "Sent out to explain the purchase of a London estate to a foreigner," the solicitor's clerk finds himself teaching Dracula the principles of his profession. "We solicitors ha[ve] a system of agency one for the other," he tells him, "so that local work [can] be done locally on instruction from any solicitor, so that the client, simply by placing himself in the hands of one man, [can] have his wishes carried out by him without further trouble" (21, 36). Dracula quickly learns how to use this network, which will function for anyone who has enough money. Consequently, the narrative includes the Count's solicitor's letter describing shipment of his boxes from Whitby to Carfax via the Great Northern Railway.

In such a network, a person at any point can produce an effect at any other simply by cabling an order. As T. A. McGraw had proposed in 1875, one could use the nervous system of another to carry out one's own desires. In the British network, the vampire recognizes a system similar to his own, and Harker reflects that Dracula "would have made a wonderful solicitor, for there was nothing that he did not think of or foresee" (37). Dracula's unpaid agents are rats, wolves, madmen, women, and gypsies, and his orders are direct mental summons, but the telegraphic and telepathic networks work the same way.

The goal of telegraphy is "transmitting intelligence," and both Dracula and his pursuers want to transmit information centripetally: from all points of their networks to themselves at the center. Because knowledge means the potential to control, each side tries to learn as much as possible about the other. As Henry James showed in his story of telegraphy, English combines "science" and "*connaissance*" in a single word, "knowledge," and Stoker's novel explores the full range of its meanings. The hunters seek science, or information, and to defeat Dracula, they bring into play an entire "international information industry" (Gagnier 147).[19] Dracula, in contrast, desires *connaissance,* knowledge in the physical and sexual sense of the word.[20] In the battle for knowledge, the hunters' papers become the equivalent of their bodies, of their nerves and blood.

Offering her diary to Seward, Mina tells him, "when you have read those papers . . . you will know me better" (196). To be acquainted, she implies, is to be informed, and when Mina finally encounters Dracula in the flesh, she "[knows] him at once from the description of the others" (251). When Lucy dies, Van Helsing asks that they seal her papers, warning, "it is not well that her very thoughts go into the hands of strangers" (148). Equating documents with thoughts, he suggests that uncontrolled access is a kind of promiscuity. No one must "know" Lucy outside of their polyandrous group. Shortly thereafter, Van Helsing asks to read all of Lucy's papers himself. Warning Seward never to share his knowledge too freely, the psychologist advises him to "keep knowledge in its place, where it may rest—where it may gather its kind around it and breed" (111). While this "knowledge" can be acquired and transmitted on paper, his phrasing betrays its close affinity to corporeal "knowledge." Like sexual energy, it must be protected and conserved so that it may survive and reproduce itself.

Although the hunters value knowledge, they have no qualms about deceiving others. They fake death certificates, and they carefully avoid the police. Observing their actions, one concludes that they respect knowledge only as a means to an end. They want to know the truth not for its own sake but because being informed makes them more powerful. As a scientist, Seward tries to make himself "master of all the facts of [Renfield's] hallucination," and Van Helsing tells him that once he reads Mina's and Harker's diaries, he will be "master of the facts" of Dracula (61, 194). These descriptions reverberate ironically with Renfield's own claim, "the master is at hand," and with Dracula's self-analysis, "I have been so long master that I would be master still—or at least that none other should be master of me" (96, 26).

Still, Stoker's characters resist seeing any affinity between their operations and Dracula's. To them, webs of connections are reserved for the forces of good alone. By showing Dracula's association with dust and decay, Stoker offers the reader the comforting possibility that evil is fragmented. The vampire's ability to dissolve himself into "quaint little specks" suggests that even his body is not as unified as organized matter should be (48). Stoker indicates Dracula's possession of the unconscious Lucy's mind with her pantomime of tearing up the letter to her protectors and "scattering the fragments" (139). Isolation and selfishness, the hunters would like to believe, are the essential qualities of evil; communal bonding and altruism those of good. Unlike *Middlemarch,* Stoker's novel never substantiates this distinction, but the characters' commitment to it reveals the ongoing moral implications of the network image.

Like Eliot, Stoker invokes the traditional notion of sympathy. From his early descriptions of theatrical performances, one can infer that he saw sympathy as a direct emotional connection. He observed a "direct force of sympathy" between the beloved actress Ellen Terry and her audience, an indication of their ability to identify with her character (Farson 50). In *Dracula,* sympathy implies an emotional exchange, the ability to feel what another person is feeling, and this quality is often—but not always—associated with women. "Though sympathy can't alter facts," writes Mina, "it can help to make them more bearable." After Lucy's death, her fiancé cries on Mina's shoulder because "there was no woman whose sympathy could be given to him" (90, 203). According to Seward, sympathy results from absolute open-mindedness, and he boasts that Van Helsing's views "are as wide as his all-embracing sympathy" (106). But if sympathy means the ability to enter another person's mind, then the most sympathetic figure in the novel is the Count. His ability to make mental connections far exceeds that of either psychologist.

In his endless praises of networking, Van Helsing lives up to Seward's description. As the hunters' patriarch and leader, the senior psychologist advocates cooperative action, arguing—unconvincingly—that their communal bonding and sense of higher purpose will give them the edge over their selfish opponent.[21] Van Helsing rallies his troops with the promise that "we are all more strong together" (274). He predicts victory because "we have self-devotion in a cause, and an end to achieve which is not a selfish one" (210). In contrast, he believes, Dracula has only a "child-brain . . . that grow not yet to our stature, and that do only work selfish and therefore small" (294). Observing the patient Renfield, Seward echoes his senior colleague's view, reflecting that "when self is the fixed point the centripetal force is balanced with the centrifugal; when duty, a cause, etc., is the fixed point, the latter force is paramount" (62). His description reflects Eliot's vision of the individual as a point on whom many forces come to bear. Selfishness keeps people from moving and acting, but altruists move outward to engage others—giving them the advantage in battle.

While these claims about Dracula's mental inferiority encourage the hunters, the Count hardly lives up to their low expectations. Repeatedly, he outsmarts the "adults" and benefits from their mistakes. Despite his child-brain's purported limitations, it is sophisticated enough to destroy their written records—except for one copy locked in a safe. In their greatest error, the hunters leave their intelligence center unprotected. Mina hates to be alone and matches Van Helsing in her love of cooperative action. When she feels obligated to act "brave and cheerful" for her husband, she complains to Lucy, "I have no one here that I can confide in" (143). Like the psychologist, she asserts that "working together and with absolute trust, we can surely be stronger than if some of us were in the dark," but the men at first fail to include her in their collective body (197). When they cease confiding in her and abandon her to hunt the vampire, the sympathetic Count accurately reads her repressed anger. When she is alone and unprotected, Dracula makes it his business to know the body that coordinates his pursuers' knowledge.[22]

The hunters recover from this attack only with great difficulty, having learned that shutting down the flow of information to any member of their group—particularly to their computer—will cripple their operation. As they chase the parasite, they can be effective only if they act as a unified body, informed by a communal nervous system. When Mina's telegram arrives to warn the group of Dracula's proximity, Seward recalls, "we all moved out to the hall with one impulse." When the Count himself appears, Seward reports that "with a single impulse, we all advanced on him" (264, 266). These two descriptions recall Henry Booth's 1845 argument for the standardization of time, the "sublimity in the idea of a whole nation stirred by one impulse" (Simmons 346). Like the earlier writer, Stoker envisions a body that acts as a unit because it obeys a single, coordinated set of signals, and he is so taken with the image that he employs it twice in one scene. The British hunters' technology, he implies, makes them superior to the vampire because it functions like the body's own communications system. But this, of course, only increases their resemblance to Dracula.

Communicating in the East

We learn much less about Dracula's communications network than we do about his opponents' because Stoker gives the vampire almost no chance to express himself. While the Count makes a number of brief, impressive speeches, his perspective is almost completely excluded from the narrative.[23] But even though we only hear about Dracula through the hunters' biased descriptions, it is clear that he has his own organized network for transmitting intelligence.

Like the hunters, Dracula works systematically, using his own telepathic network as well as his pursuers' technological ones. Studying the wounds left by the "bloofer lady," the writer of the tabloid article suspects that "whatever animal inflicts them has a system or method of its own" (160). In Dracula's castle, Harker feels as though a "net of gloom and mystery" is "closing round" him, and his observation that the Count would have made an excellent solicitor clashes with Van Helsing's deprecating description of Dracula's mind (38). While Van Helsing is the psychologist, Harker knows the Count better, and the evidence appears to be on his side. In his invasion of England, the vampire scrupulously plans ahead and works as carefully as his pursuers. Using the British networks of solicitors, carters, and trains, Dracula "achieve[s] a certain amount of distribution" (232). When the Count's opponents break into his London house, they find a bundle of title deeds and an array of "notepaper, envelopes, and pens, and ink" arranged in an "orderly disorder" on his dusty dining table. Like the hunters, he is "creeping into knowledge experimentally" (262, 264). Van Helsing looks down on Dracula's empirical, purely inductive thinking, but as a scientist, the psychologist knows that he often learns the same way.

Stoker's description of "orderly disorder" epitomizes the disturbing relationship between his heroes and his monster. Like the characters, the author and his readers would like to regard them as opposites—as order and disorder. Instead, the vampire manifests a disorder that is full of order. The Count reproduces the hunters' order, just as they reproduce his chaos. Explaining the vampire legend, Van Helsing tells the others, "it is not the least of its terrors that this evil thing is rooted deep in all good" (213).[24] Dracula has chosen to come to England, a land "most full of promise" for him, because he senses his affinity to the British (279). Like them, he functions by networking, and he aims to extend his net by annexing the British ones and connecting them to his own.

For years, Dracula has been studying Western ways, preparing for his colonization of England. Harker encounters him lying on a sofa reading the notoriously illegible *Bradshaw's Guide,* and the solicitor's clerk sees that the Count's library contains "a vast number of English books" (25). Dracula not only subscribes to British newspapers and magazines; he has them bound into volumes, reflecting the hunters' own collating process. Like his opponents, he has assembled a wealth of information, including the London Directory; Whitaker's Almanac; and lists of governmental, legal, and military personnel. In the Transylvanian castle, Harker encounters books on "history, geography, politics, political economy, botany, geology, law—all relating to England and English life and customs and manners" (25). Dracula sees easily how he can best damage his opponents, and Harker—whose opinion of the Count's abilities is consistently more accurate than Van Helsing's—observes, "he seemed to have been prepared for every obstacle which might be placed by accident in the way of his intentions being carried out" (200). Although Dracula fears that he is merely "book-smart," and the hunters try to convince themselves that he is childishly imitating their ways, he appears to have mastered them admirably.[25]

While Van Helsing and his cohorts take great pride in their intelligence network, they, like Dracula, recognize its vulnerability. Deep down, they know he can infiltrate this network as he has infiltrated England, using it for his own purposes. "What more may he not do," worries Van Helsing, "when the greater world of thought is open to him?" (279). In his use of professional and communications networks, Dracula can "pass" as British (Arata 639).

With great uneasiness, the hunters hint at their resemblance to Dracula, and they loudly assert their intellectual superiority. Harker is appalled when Dracula goes out marauding in Harker's British clothes—which fit the vampire perfectly—thinking, "any wickedness which he may do shall by the local people be attributed to me" (47). Both Van Helsing and Harker have a special affinity to the Count, the foreign psychologist echoing Dracula's threat, "I warn you that you do not thwart me" (121). When Van Helsing tells Seward about the crimes the undead Lucy has committed, Seward finds it easier to believe that his mentor has become "unhinged" and committed them himself than to accept that there are vampires running loose in England (181).

Through his depiction of Renfield ("the sanest lunatic," declares Quincey Morris, "that I ever saw"), Stoker problematizes the distinction between sanity and madness, and the haziness of this boundary makes the hunters' resemblance to Dracula all the more plausible (215). Like Mina, Renfield becomes the focus of communications, telepathic and telegraphic. Both the vampire and the hunters try to study and control him, and his affinity to both camps reveals their affinity to each other. Seward's description of Renfield as "my own pet lunatic" reflects his frequent fears that "we must all be mad and . . . we shall all wake to sanity in strait-waistcoats" (206, 240). The lunatic and the monster differ from the hunters neither in their desires, nor in their methods, nor in their gory deeds; Dracula is simply more willing to act on his erotic and violent urges.[26]

The Count, however, remains distinct from his hunters; if he did not, there would be no story to tell. While he adopts many of his British opponents' methods, their paper and wire communications never fully replace his own strange ways of communicating. What makes Dracula so terrifying is his *choice* to be different. Like an unruly native, he has observed Western ways and adopted those that will benefit him, but for the most part, he has retained his own.

Dracula differs most drastically from the hunters in his relationship to time. Because of his immortality, the Westerners' quest for linear progress does not move him. The Count thinks in terms of centuries, not hours, minutes, and days.[27] Cycles of daylight mean more to him than fears of a remote death, for it is these that most restrict his actions. Seeing the Englishmen's preoccupation with time, he tries to demoralize them by convincing them that time works in his favor and is beyond their control. To disorient Harker, the Count prevents him—by hypnotic suggestion—from winding his watch. The young solicitor, a creature of habit, awakens to find it unwound and writes defensively, "I am rigorously accustomed to wind it the last thing before going to bed" (44). When Dracula invades England, he uses a similar strategy, trying to prove to his enemies that "time is on my side" (267). Once the Count has infected Mina, Van Helsing admits that "time is now to be dreaded," and the American Morris, too, concedes that "time is everything with him" (273, 214). Dracula's reading of Bradshaw has taught him to control the timetable of their struggle, which he would very likely have won had he not been subject to the timetable of the earth.

Despite his immortality, Dracula is restricted by natural cycles, and Stoker combines this fact of vampire lore with a more general belief that evil moves in circles. The "good" hunters try to progress linearly, but Dracula, in Stoker's descriptions, exerts his force in disorienting waves. As Dracula drives Harker toward his castle, the young solicitor senses that they are circling—"that we were simply going over and over the same ground again" (18). The Count rules over a land that is a "whirlpool of European races" and an "imaginative whirlpool" of "every known superstition," so that his longing for the "whirl and rush of humanity" in London is easy to understand (33, 10, 26). When Dracula invades Mina's mind, she reports that "things began to whirl through my brain just as the cloudy column was now whirling in the room" (227). Evil, Stoker implies, robs its victims of their progressive, linear perspective, with regard to space as well as time.

Dracula's colonial strategy, too, is based on circular patterns. Picturing Dracula's invasion of England, Harker fears that he will "create a new and ever-widening circle of semi-demons to batten on the helpless" (53–54). In terms of population genetics, Dracula's creation of a new, undead race is a highly complex function with an unpredictable output. Van Helsing, too, sees it as a spreading circle of death. Once the vampiric pebble has been thrown into the sea of England, he believes, "so the circle goes on ever widening, like as the ripples from a stone thrown in the water" (190). Because each new vampire can create many others, the undead become a network extending itself in every direction.

Despite the hunters' advocacy of linear progress, Dracula's network of concentric circles mirrors their own web. But while his system transmits information very efficiently, it also differs markedly from the British net. In the vampiric network, as in Diderot's web, all information flows to and from a central point. Dracula gathers intelligence and keeps himself informed, but his minions do not communicate with each other. In the hunters' web, all data flows through the central processor, Mina, but she merely informs, reasons, and advises, letting the hunters make the command decisions. Unlike the typist, Dracula is a motor neuron as well as an interneuron. Whereas the hunters' power is distributed dynamically among different points in the web, his remains fixed at the center.

Like Diderot's spider, Dracula uses his network of nerves to inform himself of dangers and issue orders, but his wireless web gives him access to the networks of other organisms. As T. A. McGraw described, he exploits other creatures' nervous systems to perform tasks for his own benefit, operating them by remote control. With his telepathic and telekinetic powers, he can "command all the meaner things," most notably rats and wolves (209). Yet while he can summon almost any living creature, his control is never complete. He cannot overcome a rat's fear of a terrier or—the hunters' hope—a woman's will to be good. From Emily Gerard's "Transylvanian Superstitions" (1885), Stoker learned about Eastern European gypsies, "whose ambulating caravans cover the country as with a network" (Stoker, "Notes for Dracula" 332). In *Dracula,* these gypsies figure as one of the meaner things the Count commands—loyal as long as his gold holds out and as subject to fear as the rats. Dracula succeeds quite well, however, in his communications with British women, whose civilization has repressed their anger and their desires. Just as Dracula successfully uses Western networks, the hunters triumph only by adopting Dracula's methods of communication—by jacking into a woman's unconscious mind.

Telegraphy vs. Telepathy

As leader of the vampire hunters, Stoker's character Van Helsing demands open-mindedness in every possible sense. Like the scientists of the Society for Psychical Research, he uses the astonishingly rapid development of communications technology to argue that mental phenomena dismissed by contemporary psychologists will someday be accepted as reality. "It is the fault of our science that it wants to explain all," he tells Seward, "and if it explain not, then it says there is nothing to explain" (171). Beginning with the least plausible, least accepted psychical phenomena, Van Helsing presents Seward with a continuum to see where his belief begins. "I suppose now you do not believe in corporeal transference. No? Nor in materialization? No? Nor in astral bodies. No? Nor in the reading of thought. No? Nor in hypnotism—" Here Seward interrupts him, for he does believe in hypnotism. "Charcot has proved that pretty well," he says (171). Echoing the arguments of Barrett, Gurney, and Myers, his teacher then asks him how he can accept hypnotism when he rejects thought transference.

For Van Helsing, all of these psychical phenomena involve the transmission of mental energy, and it is ignorant to impose boundaries on such a continuum. He believes in Dracula's powers for very "scientific" reasons. Knowing that electricity can be used to manipulate living organisms, he tells Seward, "there are things done today in electrical science which would have been deemed unholy by the very men who discovered electricity" (171).[28] For almost three decades, physiologists had known that one could produce specific movements by electrically stimulating particular areas of the brain, and it seemed reasonable to expect that one might do the same using remote signals (Greenway 224–25).

Like members of the Society for Psychical Research, Van Helsing bases his argument for telepathy on the accepted idea of interrelated forms of energy. Dracula's homeland, he explains to Seward, is "full of strangeness of the geologic and chemical world, . . . [and] doubtless, there is something magnetic or electric in some of these combinations of occult forces" (278). To communicate, he implies, the vampire uses the same forces that the hunters do: pulses of energy that affect the mind as they do needles and wires. Both the body and the telegraph rely on electrical signals, and while Westerners have adopted the technological system for long-distance transmissions, Dracula has adopted the organic one.[29]

Like Twain, Stoker was deeply interested in psychical phenomena, and his busy social life kept him apprised of new discoveries and trends. Stoker concentrated in science at Trinity, and he remained close to his brother William, a successful surgeon and inspector for the Vivisection Act (Farson 228). Stoker was especially intrigued by studies that explained behavior physiologically, including the work of Franz Mesmer (1734–1815), Charles Bell (1774–1842), Charles Darwin, and Herbert Spencer (Glover 988–89). Like many other late Victorians, he wondered about the body's ability to respond to stimuli that did not register consciously, and like Barrett and Myers, he suspected that if people could communicate telepathically, it was their unconscious minds that made contact. Most scientists did not believe that telepathy could be studied scientifically—if, indeed, it existed at all. In Van Helsing's exchange with Seward, the more "open-minded" public—cast as a senior psychologist—lectures scientists on science. Stoker's novel supports the possibility of telepathy, for the scientists' disbelief in Dracula's powers is the vampire's "greatest advantage" (Feimer 167).

Hypnotism fascinated Stoker, just as it intrigued late nineteenth-century psychologists and psychical researchers, because it demonstrated influence that circumvented conscious control. In *Dracula,* he presents the mind as a sensitive receiving device whose openness can be as great a danger as an advantage. Like most psychologists of the 1890s, Stoker believed that the brain normally maintains defenses, for it has been carefully closed off by cultural teachings. Van Helsing explains that he has had to "train [himself] to keep an open mind" (209). In an earlier story, Stoker wrote that "Uneasiness is an instinct and means warning. The psychic faculties are often the sentries of the intellect, and when they sound alarm the reason begins to act, although perhaps not consciously" (Farson 100). Stoker's novel supports this model of the mind, for Dracula succeeds by contacting the unconscious directly, slipping in while the sentries are asleep.

Like many scientists, Stoker mistrusted the mind's defenses, and he feared the power of suggestion. In an article of 1908, he argued for literary censorship, urging that society supplement the mind's own watchful strategies:

> In all things of which suggestion is a part there is a possible element of evil. Even in imagination, of whose products the best known and most potent is perhaps fiction, there is a danger of corruption. . . . the only emotions which in the long run harm are those arising from sex impulses, and when we have realized this we have put a finger on the actual point of danger. (Farson 209)

By the time he wrote this warning, Stoker would have been aware of his imminent death from tertiary syphilis, so that he would have been painfully conscious of what could enter the body. For writers of this period, this deadly disease served as a metaphor for equally deadly ideas that could penetrate and corrupt a mind. In *Dracula,* written more than a decade earlier, Stoker already seems preoccupied with the mind's inability to screen out dangerous signals.

Supposedly, some minds are more open than others, and Stoker follows most Victorian scientists in presenting the female as more susceptible than the male. Describing an early encounter with his idol Henry Irving, Stoker emphasized the actor's "magnetism" and "commanding force," going to great lengths to explain why he became hysterical one evening when Irving recited a passionate poem. He was "no weak individual," Stoker insisted, "yielding to a superior emotional force" (Farson 28, 30). In his novel, Stoker attempts—unsuccessfully—to present such nervous sensitivity as a female characteristic. It is with British women that the magnetic vampire has intercourse.

In *Dracula*, mental communication is predominantly heterosexual. The Count's "awful women" try to hypnotize Harker, while Dracula himself inserts his suggestions into the hunters' mates (44). The women's natural sympathy and sensitivity, Stoker implies, leave them open to the vampire's seductive transmissions, and the men must try to protect their mates' minds just as they protect their bodies. Particularly telling is Lucy's claim that she "sympathizes" with Desdemona, who "had such a dangerous stream poured in her ear, even by a black man" (59). While she makes the comparison to describe a marriage proposal from Quincey Morris, she anticipates the approach of her real husband, the Count. Her reference to a "dangerous stream" reveals not just racial and sexual paranoia but her sense that she is being bombarded with intriguing messages, signals to which she wants to respond.

Dracula selects Lucy for her appetite—her expressed desire to marry three men—but also for her high fidelity as a "receiver." According to Mina, "Lucy is so sweet and sensitive that she feels influences more acutely than other people do" (85). When her defenses are down, in her dreams and sleep, she is as open to the Count as any telegraph key. "All this weakness comes to me in sleep," she confesses to Seward, expressing the Victorian fear that unconsciousness offers free access to the unprotected mind (116). As Barrett and Myers would have predicted, Dracula communicates directly with her unconscious. As his control increases, the reader can follow the struggle between her animal impulses and the reasoning will that normally represses them. When Lucy's shields are lowered, her sleepwalking takes her straight to the Count, who has summoned her by mental telegram. When she is conscious and rational, she fears sleep. She pulls the garlic flowers up against her throat and protectively clutches her letter to her rescuers. Later, in her "lethargic state," Lucy tears up the letter and pushes the flowers away. Stoker shows Dracula's final triumph by having Lucy express the most harmful of all emotions, her sexual impulses, voluptuously begging her fiancé for a kiss.

But how has Dracula reduced a respectable woman to such a state? Stoker denies the vampire the opportunity to describe his own communications, but we learn of his techniques through the other characters' descriptions. When Dracula approaches Mina and Lucy, they both report a feeling of paralysis. Lucy writes in her final memorandum, "I tried to stir, but there was some spell upon me" (131). Mina, too, explains that "I was powerless to act; my feet, and my hands, and my brain were weighted" (227). Like a fiendish physiologist, Dracula has taken control of their motor systems, overriding their wills with his more powerful electromagnetic impulses. Or perhaps, like a skilled hypnotist, he has merely convinced some higher center that he has the motor system under control.

Like percipients interviewed by the Society for Psychical Research, Dracula's victims receive mainly visual impressions. When the Count transmits suggestions, he does so with images, not words, leaving his correspondents to interpret them as they would any dream. "He began promising me things—" describes Renfield, "not in words but by doing them" (244). By showing the mental patient millions of rats, Dracula conveys his offer of countless lives. Dracula's telepathic network uses the code of dream images, and because the unconscious mind receives his signals directly, his system is enviably efficient.

When Van Helsing hypnotizes Mina, his mental communication comes across as a pitiful imitation of Dracula's, and in their efforts to read and transmit thoughts, both Van Helsing and Seward are "wanna-be" vampires.[30] According to Lucy, Seward is always "looking one straight in the face, as if trying to read one's thoughts," and Seward himself writes frustratedly, "I wish I could get some glimpse of [Renfield's] mind" (57, 110). Van Helsing successfully opens a line of communication to Mina's unconscious, but he is no match for Dracula, who simply "sen[ds] her his spirit to read her mind" (294).

From the time that she first meets Seward, Mina acts as the central processor in the hunters' communications net, receiving, transmitting, and organizing their information. When Dracula annexes her, incorporating her into his wireless web, her mind and body become the crossover point between two analogous but very different intelligence systems.[31] Once a line of communication has been opened, the hunters learn, it can transmit information in both directions. Mina's mental contact with Dracula allows them to track him by jacking into his perceptions of his journey, but they realize that any strategies they reveal to Mina will reach Dracula via the same link.

Like Twain, Stoker draws on his knowledge of mesmerism and spiritualism when he represents long-distance communications. Alison Winter has shown that from the 1840s onward, the mesmerized or entranced woman enjoyed considerable authority as an "oracular figure" or "prophetess." In séances, the spiritual medium offered her mind and body as passive vehicles for the flow of data. "Rendered machinelike," she supplied listeners with otherwise inaccessible information (Winter, *Mesmerized* 227). The image of the female medium providing "reports" was a common one in Victorian culture, and Stoker incorporated it into his descriptions of the hunters' computer, Mina.

As Seward records Mina's "hypnotic reports," the parallel between telepathy and telegraphy becomes increasingly apparent. Sentence for sentence, Stoker alternates these reports with references to telegrams from London, which, like the hypnotic briefings, inform them about Dracula's position. Seward writes in his diary: "Mrs. Harker reported last night and this morning as usual: 'lapping waves and rushing water,' though she added that 'the waves were very faint.' The telegrams from London have been the same: 'no further report'" (292). It is only by relying on both networks—the organic and the technological—that he gains the necessary information.

Unlike the wires, Mina's mind can resist communicating, and as Dracula takes it over, he gradually alters her circuitry so that the hunters' central processor will only transmit one way. Despite his sabotage, her living circuits can accomplish things that the wires cannot. In an inspired set of deductions, she predicts the route that the Count will take back to his castle, allowing the hunters to ambush and destroy him. Like Sherlock Holmes, she uses her intimate knowledge of the invader to predict his behavior.

But *Dracula* depicts no triumph of telegraphy over telepathy, nor does it depict the human mind's triumph over the machine. One contemporaneous reviewer wrote that "the up-to-dateness of the book—the phonograph diaries, typewriters, and so on—hardly fits in with the medieval methods which ultimately secure the victory for Count Dracula's foes" (365). Stoker's reliance on cutting-edge communications technology, though, only highlights bodies' resemblances to machines. As Stoker represents them, corporeal and technological communications are inseparable, and it is only a particular combination of them that leads to the Count's defeat. In *Dracula*, telepathy resembles telegraphy as closely as evil resembles good.

Parasites

The movement of information is only one type of circulation that Stoker's novel depicts. Like *Middlemarch*, *Dracula* also traces the movements of blood, money, and energy, all of which are related to each other and the flow of data. As Stoker depicts them, however, none of these fluxes is a true circulation, for nothing in *Dracula* really moves in a circle. Like information, blood and energy flow one way: toward the parasites, toward those in positions of power.

Michel Serres has described the parasite in terms that reveal the close affinity of physiology, information theory, economics, and thermodynamics, four different kinds of "communication." Irreversible flow, he argues, is the most common of all relations, not simply in biology but in most complex systems (16). A parasite, which in French can refer to a noise in a circuit, introduces a new order to a system by disrupting its current relations; it produces even as it consumes (Serres 9). A parasite does not live off of anyone directly, but off of the relations between communicating beings. Surviving through mimicry, it is the consummate hypocrite, imitating those around it to avoid rejection (Serres 272). When a parasite looks for relationships off of which to feed, it seeks the unusual, hoping for opportunities to enter the relation (Serres 201). Cybernetically, parasitism means the theft of information, the blood whose flow allows host and parasite to survive (Serres 57). Once it has incorporated itself into a system, the parasite commands considerable power, receiving everything and giving nothing. As Serres puts it, "he whose only function is to eat will command."[32]

Thief, seducer, creator, and mimic, Dracula is the ultimate parasite. The whole point of vampirism is sucking other people's blood—living at other people's expense. A vampiric novel, *Dracula* is itself parasitic, draining the reader's energy and reproducing itself in the reader's excited accounts of it.[33] One of the novel's most terrifying moments comes when Harker sees Dracula for what he really is: an "awful creature . . . simply gorged with blood . . . a filthy leech, exhausted with his repletion" (53). He trembles at the thought of "those awful women who were—who are—waiting to suck my blood" (44). Coming from a region in which "hardly a foot of soil . . . has not been enriched by the blood of men," Dracula and his women become identified with irreversible flow (27). As long as the Count moves freely through Britain, the nation's blood drips slowly into a funnel: from the hunters to Lucy and then on into Dracula's mouth.

When the hunters try to replenish Lucy with transfusions, Stoker's descriptions reveal the broader implications of bleeding. After her fiancé revives her with his blood, Lucy writes that "somehow Arthur feels very, very close to me. I seem to feel his presence warm about me" (117). Her perception suggests a connection between the flow of blood and the flow of thoughts, as though she had acquired something of his mind along with his corpuscles. A leaky vessel, Lucy absorbs the blood of Holmwood, Seward, Morris, and finally even the aged Van Helsing. The practical Morris observes that in ten days' time, "that poor pretty creature that we all love has had put into her veins . . . the blood of four strong men. Man alive, her whole body wouldn't hold it. . . . What took it out?" (138). Because she is parasitized in turn, Lucy can drain their lifeblood indefinitely. It is Dracula who is the bottomless pit, with an infinite capacity to absorb.

Stoker's characters speculate that sharing blood implies a sexual merging, making Lucy "truly [Arthur's] bride." By this standard, however, Lucy is a "polyandrist," and she and Mina are the brides of Dracula—as are all the hunters whose blood has flowed through Lucy (158). When Stoker describes Mina's son in the final note, the boy emerges as the endpoint of all blood, information, and energy. Quincey Morris has died in the final battle with Dracula, and Mina holds a "secret belief that some of our brave friend's spirit passed into [the boy]." Her newborn son bears a "bundle of names [that] links all our little band of men together," fathers whose information flowed into Mina and made the conception possible (326). If this is the criterion for fatherhood, however, then the boy should also bear the name of Dracula, whose thoughts poured into Mina's mind as his blood gushed into her mouth.

Stoker's descriptions of Renfield's mania also tie the trickling of blood to the flow of information because they identify it with the transfer of energy. Adopting a Victorian advertising slogan, "The Blood Is the Life," the lunatic Renfield turns mass culture's exploitation of religion back into a religion again, of sorts (206–7). Envisioning life as a food pyramid or cascade, he makes it his business to "absorb as many lives as he can" (71). Like Mina, he accepts ideas both from the rational Seward and from the bloodthirsty Count, consuming flies, spiders, and a sparrow but recording his feats carefully in a little book. In a revealing remark, Seward senses that Renfield is "mixed up with the Count in an indexy kind of way" (219). He means to say that the lunatic is an informer, but his choice of words emphasizes Renfield's preoccupation with lists. The madman is an index that points both to the hunters' thoughts and to Dracula's.

With his delusions of grandeur, Renfield imagines himself at the crux of a food pyramid, eating animals that have nourished themselves with countless other lives. He is not alone in the novel in his hunger to consume: Dracula is always hungry, and after being reminded that Dracula has "banqueted heavily" on her, Mina tells the hunters, "we must all eat that we may be strong" (258). These ongoing references to bleeding and eating reflect the nineteenth-century tendency to perceive physiological and mental activity in terms of the conservation of energy. Following Helmholtz's 1847 claim that energy was preserved when it changed its forms, physiologists proposed that "one acquires the energy of what one consumes" (Greenway 217). Because of these cultural associations, Dracula's parasitism emerges as a thermodynamic as well as a biological threat.

Stoker's representation of the Count's bloodsucking brings to life Victorian fears of an energy drain, a degeneration of the nation's vital force (Arata 622). While the hunters' narrative often refers to their strength, the "brave men" repeatedly fail to deliver. It is Dracula whose body is strongest and Mina whose mind proves most powerful. By blaming Dracula for the energy crisis, Stoker sidesteps the more disturbing possibility of an inevitable decline suggested by the second law of thermodynamics. Bled by a foreigner who reproduces himself infinitely, the British men are barely able to produce the next generation.

To Victorians, the flow of blood implied the transmission of family traits, family names, and—most importantly—family fortunes, so that biological and economic flow occurred through common channels. Besides controlling the movement of blood through much of the novel, Dracula commands considerable wealth; even with his physical vigor and formidable intelligence, he could never have colonized England if he were not also a very rich vampire.[34] When Harker slashes the Count, a "stream of gold" falls out of the vampire's coat, suggesting that blood means money (Stoker, *Dracula* 266; Halberstam 348). In Dracula's castle, Harker spots "a great heap of gold . . . covered with a film of dust, as though it had lain long in the ground" (50). In the bowels of Dracula's castle, Harker is seeing dead money, money that has been taken out of circulation, and his disgust at the smell of death includes his disapproval of hoarding.[35]

Dracula, however, is restoring his capital to general circulation just as he is introducing his body to new circuits. As an aristocrat who spends money and travels on trains, he would seem to please middle-class businessmen who upheld circulation as a moral value (Schivelbusch 195). Dracula uses money to acquire information and information to acquire blood, but we never learn how he has acquired his money. While these forms of energy are interconvertible, there is no true circulation in Stoker's novel, only expenditure.

Like Eliot, Stoker depicts the interdependence of many forms of communication: the movement of knowledge, the transport of people, and the flow of money. Written a quarter century later than *Middlemarch*, *Dracula* takes a much more ambiguous stance toward these types of communication. Both the Count and his pursuers rely on the free flow of traffic: the vampire can occupy a London home and the hunters can burglarize it because "in the very vast of the traffic there is none to notice" (255). In an open communications system, one can hunt down intruders, but parasites can easily insinuate themselves into the relations of everyday life. Whereas Eliot celebrates the dynamism of communications networks, Stoker stresses their potential for exploitation. In his novel, it is the parasite, not the web, that is evil, but England's formidable network of trains, telegraphs, and solicitors holds the potential to do ill because it can be commanded by anyone. Future battles between good and evil, Stoker suggests, will be struggles to control information.

Conclusion: Wired Thoughts

> In the dense entangled street,
> Where the web of Trade is weaving,
> Forms unknown in crowds I meet
> Much of each and all believing;
> Each his small designs achieving
> Hurries on with restless feet,
> While, through Fancy's power deceiving,
> *Self* in every form I greet. (Maxwell 593–94)

As a means of representing communication, the web is not a new image. In "Reflex Musings: Reflections from Various Surfaces," James Clerk Maxwell uses a web metaphor to describe the interactions involved in economic exchanges. Like Eliot, he warns against the human tendency to project oneself outward onto others. In a network of relations, though, where does one's identity end?

When we ask "what's new about networking," writes Allucquère Rosanne Stone, we encounter two immediate responses: either everything or nothing. Either our web of computers provides a novel medium for "dramatic interaction," or it provides the same kind of opportunities for communicating as did nineteenth-century "network prostheses" (15–16). As Maxwell's poem shows, one does not need computers to envision webs. The young physicist may have selected the web image because he was pondering Faraday's lines of force, or because he was using England's growing telegraph network, or maybe simply because his culture had adopted it as a way to describe social relations. Then, as now, the metaphor invited all who encountered it to rethink their identities. What is new about networking is very little, and examining nineteenth-century understandings of communications webs helps to ground our own explorations of an interconnected world.

Developed by scientists who studied organic "receivers," the telegraph altered the way people viewed their own bodies. Telegraph operators perceived their keys and wires as extensions of their hands and minds. At times, their hands and minds seemed like extensions of their wires and keys. The fact that so many telegraphers were female suggests a consensus among those who controlled early networks that communication was women's work. Donna Haraway has proposed that the female body is "coded as a body-in-connection," the male as a "body-in-isolation" (Balsamo 136). From the 1850s onward, as more and more people began communicating through telegrams, the public—male and female—began to understand themselves as "connected" and to envision themselves as cross-points in a net.

Traditionally, observes N. Katherine Hayles, personal boundaries have depended on bodily membranes, coinciding with the surface of the skin. But "if our body surfaces are membranes through which information flows," she asks, then *"who are we?"* (*How We Became Posthuman* 109). In a network, individuals—if there are individuals—exist as selective transmitters rather than as bounded cells, defined more by relations than intrinsic traits. Allucquère Rosanne Stone recalls an attempt to hear Stephen Hawking "speak" with the aid of a computerized voice generator. Dissatisfied with her position "outside" on a lawn, where she was listening to the physicist through a PA system, Stone pushed her way "inside," to the very front row of the auditorium, so that she was only meters away from Hawking's body. But then, watching the brilliant scholar whose organic communications system made it impossible for him to convey thoughts without a computer, she began to wonder, "exactly where *is* Hawking?" (4–5). Except for those who knew the physicist before the onset of his amyotrophic lateral sclerosis, most people who think of Hawking immediately envision his wheelchair and voice box. Together, the body, mind, and technology form a unified image that represents the technologized subject we know as "Hawking."

While Hawking presents an extreme case of merged organo-technical communications systems, anyone who has ever used a computer can attest to the exhilaration of receiving E-mail from Africa or the frustration of being told a connection has been severed. Both the popular and the "high" literature of the nineteenth century depict this same investment of identity in technology, describing telegraphers' joyful feelings of power along with their bitter acknowledgment of their own limitations. Then as now, electronic prostheses promised that one's body might extend as far as one's signals could be transmitted. But does a technological network really offer the same kinds of communication provided by a body?

On the one hand, electronic communication seems to offer certain advantages over physical intimacy. As early as 1846, American professor Alonzo Jackman speculated that the telegraph was "useful for prophylaxis," permitting communication far less risky than physical contact (Marvin 201). In the era of cybersex, we have found that technological communications transmit only certain kinds of viruses. Electronic eroticism offers intimacy while ensuring that one's bodily membranes will not be permeated. Why then do we feel violated when we learn that someone has broken into our computer accounts?

The sense of panic one experiences on hearing that one has fallen victim to a hacker may exceed the outrage one feels on learning that one's house has been burglarized. Files are more than possessions, and knowing that one has been violated cybernetically is like knowing that someone has broken into one's head and rifled through the thoughts in one's mind. Allucquère Rosanne Stone believes that out of respect for those who have been physically attacked, the word "rape" should not be extended to unwanted cybernetic penetrations (172–73). While arguing that computers have changed our understanding of presence, she maintains that the invasion of a computer account is one thing, the invasion of a body quite another.

But the more people depend upon computers to exchange thoughts, the more they identify their minds and bodies with the machines that assist them. People who are "unhooked" from the mass media report feeling "slightly less alive" (Carey 1). Into our notions of selfhood, we have incorporated the machines that bring us news about the world. It seems a reasonable move, since like our nerves, they provide us with essential information.

Since the nineteenth century those who have used electronic media to communicate have noticed that wires deliver only certain kinds of information. When Norbert Wiener developed the science of cybernetics, he decided that "communication is about relation, not essence," a fact evident to every telegrapher who ever struggled to hear pauses between letters and words (Hayles, *How We Became Posthuman* 91). Over the wires, messages assume meaning only in relation to other messages, and one can distinguish "John is dead beat, home at three" from "John is dead, be at home at three" only if one knows the sender's usual style. Like the nervous system, communications technologies inform us of patterns, not essences (Hayles, *How We Became Posthuman* 98).

Isolated from the vocal intonations and facial expressions that accompany speech, wired thoughts lack human context. In cyberspace, observes one "net surfer" quoted by Mark Dery, "everyone speaks with flattened affect." Considering why on-line exchanges so easily escalate into vicious "flame wars," Dery speculates that their abstraction from bodily presence may be removing inhibitions. "Emoticons"—visual symbols for facial expressions—provide little help and are merely a pitiful attempt to "telegraph facial expressions" (Dery 2). When thoughts are wired, misunderstandings are frequent. Failing to perceive intended meanings, correspondents project their own meanings onto disembodied words. Like telegraphy, E-mail lets us read without seeing.

In 1883, Sarah Orten answered a personal ad in the *Cincinnati Enquirer* and took a liking to the gentleman who had placed it. After the two had corresponded for some time, they were married by telegraph and remained so until Orten met her husband face to face. Much to her horror, she discovered that he was "a colored man" (Marvin 93–94). Satisfied with his words and with the persona he had presented, she balked only when faced with information their electronic coupling had not revealed. His letters had never mentioned the color of his skin, with all the associations their culture attached to it.

The concept of long-distance communication, of course, is as old as writing itself. Anyone who has exchanged letters knows how deceptive messages can be once they are released from their senders. Unaccompanied by intonations or facial expressions, they are open to any reading. Even worse, as Choderlos de Laclos illustrated in *Liaisons Dangereuses,* they are often intended to deceive. Telegraphy and E-mail differ from traditional letter-writing in that transmission is simultaneous. With no delay to prepare thoughts or anticipate answers, the illusion of real conversation becomes much more powerful. One seems to receive thoughts as quickly as they are formulated, so that one might as well be talking face to face. Because of this simultaneity, one can "talk" to a telegraph key or computer as one would to a friend.

Through interactions like these, technologies gradually alter our language and thought. Consequently, they destabilize the identities we have constructed through language. Donna Haraway describes E-mail as "passage points . . . through which identities ebb and flow" (4). As can be seen in the stories of nineteenth-century telegraphers, communications technologies invite people to try out alternative personae. "The nets are spaces of transformation," reflects Allucquère Rosanne Stone, "identity factories in which bodies are meaning machines" (180).

Yet while our communications systems have developed enormously in their capacities to transmit information, they have inspired much of this experimentation because of what they *fail* to transmit. In the early 1950s, cyberneticist A. M. Turing proposed an experiment in which subjects communicating solely by computer had to determine whether they were "talking" to a man or a woman. In a second set of experiments, they had to distinguish a human being from a machine (Hayles, *How We Became Posthuman* xi). While we have retained some of the quandaries of nineteenth-century hackers, we must now worry not just about who is at the other end of the wire but about whether there is anyone there at all.

In "The Machine Stops" (1909), E. M. Forster voiced this fear, challenging comparisons of electronic and physical communication. Depicting individuals isolated in underground cells, he suggested how their reliance on a machine—not just for communication but for all their physical needs—undermined their ability to form relationships (Perkowitz 2). While Forster's fictional internet seems to put people in touch, it actually discourages them from touching. It is no continuation of its subjects' nerves; there is nothing corporeal about it. Designed to enhance the body's communications net, it deprives people of their natural abilities to communicate. When the machine breaks down, society breaks down with it, sapped by a technology intended to maintain social bonds (Perkowitz 2).

As the network image has flickered through scientific and literary texts in the past two centuries, it has carried varying connotations and has provided a "natural" model to affirm strikingly different visions of the individual and the state. It would be wrong to classify these contrasting uses chronologically, for there has been no smooth transition from one to another, only a tug-of-war as groups with different interests have fought to appropriate an appealing image. As an image equally applicable to bodies and technologies, the web legitimized social structures by making them seem organic. At the same time, it reinscribed the body as a reflection of society. By identifying nerves with technological communications systems, writers could win sympathy for their own ways of seeing.

From the beginning, argues James W. Carey, two different understandings of communication have persisted in the West, both with religious origins. The first—transmission of information "for the purpose of control"—leads people to see society as a "network of power" (Carey 15, 34). The second, in contrast, presents communication as a force "that draws persons together in fellowship and commonality." When defined this way, communications systems aim not to impose order from a central point but to encourage local bonding that will maintain "an ordered, meaningful cultural world" (Carey 18–19). The divergent uses of network metaphors reflect these opposing definitions.

In some cases, the nerve network has been used to show the efficiency of centralized power. In 1897, Basil Williams wrote that "the [Indo-European telegraph] line seems calculated to give a strong impression of English solidity and power. . . . Indeed, no better object-lesson could be found of the superiority of English energy and enterprise over Oriental apathy and incompetence" (159). Half a century earlier, when Emil DuBois-Reymond proposed that the telegraph was "modeled" on the nervous system, he had a particular goal in mind. In a popular lecture, he used the analogy to argue that Germany's new telecommunications system was superior to France's, because it more accurately reflected the body's own web. For DuBois-Reymond, who conducted his research in Berlin, centralization seemed like a good strategy for both organic and technological communications systems. Legitimizing Werner von Siemens's telegraph network through references to organicity, the German physiologist simultaneously legitimized the expanding empire it served.

As DuBois-Reymond's analogy suggests, centralized webs "wired" empires. For European nations, centralized communications networks permitted "the transition from colonialism, where power and authority rested with the domestic governor, to imperialism, where power and authority were absorbed by the imperial capital" (Carey 212). In 1860, a writer for the popular *Cornhill Magazine* concluded his history of the telegraph by claiming:

> Already it has become an indispensable agent of civilized society—materially influencing the political, social, and commercial relations of every country in Europe. . . . We cannot but feel convinced that science, in this her most brilliant achievement, has placed in our hands an instrument which adds another link to that chain of causes which is slowly, silently, and imperceptibly bridging over the chasms which separate nation from nation and race from race; and whose influence on the future of civilization it is impossible to estimate. . . . it conveys its own significant lesson to the Indian in his wigwam, to the Hottentot in his kraal, and to the Arab in the desert. ("Electricity and the Electric Telegraph" 73)

The words "empire" and "imperialism" came into being shortly after 1870—the same years in which scientists began seeing networks in nerves.

Often, however, the network image has been used to encourage local organization that *subverts* central authority. In an early twentieth-century story by Gertrude Barnum, a union leader suggests "wiring" workers to organize them against management. "All you have to do is to wire them for electricity," observes the organizer, "keep up the current, and make the connection" (Stubbs 267). Through this metaphor, Barnum presents networking as a way to resist centralized power. When nineteenth-century telegraphers wrote about their exploits over the wires, they, too, represented their activities as a defiance of social restrictions. Rather than inhibiting them, the wires allow them to bond with others trapped like themselves.

What makes network metaphors so intriguing is that they can never be tied to a single ideology. Images offered by cultures never fully determine what people see, and the diverse uses of web imagery reveal a struggle among groups competing to control a popular representation of power. On the one hand, webs of wires or nerves convey the terrible efficiency of centralized power networks. On the other, they show the importance of local bonds in any given region, suggesting that by networking, oppressed individuals anywhere in the system can resist the will of a remote tyrant. As a cultural metaphor, the network offers interpretations of personal experiences, but personal experiences suggest new ways to understand networks.

These divergent uses of the network image confirm Michel Foucault's hypothesis that webs of power create identities; they do not crush preexisting ones. In *Discipline and Punish,* Foucault describes the Panopticon, on which many nineteenth-century Western institutions were architecturally modeled. As the structure through which social power networks exerted their force, the Panopticon distributed individuals in space, restricting their movements so that they could be watched by an unseen central observer. But even in this structure, argues Foucault, "it is not that the beautiful totality of the individual is amputated, repressed, altered by our social order, it is rather that the individual is carefully fabricated in it, according to a whole technique of forces and bodies" (217). Foucault uses the network image to show how the state exerts power over individuals but also to demonstrate how individuals respond to its forces. For Foucault, oppression and resistance are inseparable, and without the information systems that "restrict" the individual, there would be no individuals to restrict.

In the first days of the third millennium, we live as ganglia in a network of organic and technological communications devices. As Henry James suggested over a hundred years ago, we are all telegraphers, expecting unreliable bodies and machines to inform us about the world. Functioning as part of a network, however, does not mean being powerless. Unlike the unfortunate individuals assimilated by the Borg, we have not been stripped of any preexisting identity (Foucault 217). As telegraphers and physiologists discovered long ago, networks both empower and disempower. They offer exciting new relationships and relative knowledge even as they destroy obsolescent fantasies of autonomy.

Notes

1. The original German reads: "Das Wunder unserer Zeit, die elektrische Telegraphie, war daher längst in der thierischen Maschine vorgebildet. Aber die Aehnlichkeit zwischen beiden Apparaten, dem Nervensystem und dem elektrischen Telegraphen, ist noch tiefer begründet. Es ist mehr als Aehnlichkeit, es ist Verwandtschaft zwischen beiden da, Uebereinstimmung nicht allein der Wirkungen, sondern vielleicht auch der Ursachen" (*Reden* 2: 51). My translation.

2. According to American neurologist S. Weir Mitchell, who studied briefly with Bernard in 1851, when Mitchell declared innocently, "I think so and so must be the case," the physiologist asked him, "Why think when you can experiment? Exhaust experiment and then think" (Mitchell 103–4).

3. Timothy Lenoir has demonstrated that German physiologists Hermann von Helmholtz and Emil DuBois-Reymond, the nineteenth-century scientists who made the greatest contributions to neurophysiology, were in close contact with communications engineers like Werner von Siemens. By developing models that reflected the apparatus they had adapted from physics and media technology, Helmholtz and DuBois-Reymond changed people's understanding of the way nerves transmit impulses. I agree strongly with Lenoir that such comparisons between organic and technological systems were not mere devices for popularization but became incorporated into the scientists' vision and understanding of the nervous system (Lenoir, "Helmholtz" 185–88).

4. An 1860 article in the British *Cornhill Magazine* suggests how Galvani's discovery was mythologized and how lay readers understood his contributions to telegraphy and physiology:

> All the world knows the famous story of the origin of galvanism, as recorded by Arago; how, in the year 1790, Signora Galvani, the wife of a Bolognese professor, caught cold, and had frog soup prescribed for her use—how some skinned frogs lying near an electric machine, which was accidentally set in motion, gave what seemed signs of vitality, in virtue of the law of induction; and how, on passing copper hooks through their limbs, and suspending them on an iron railing, equally strong convulsions resulted, even in the absense of any apparent exciting cause ("Electricity and the Electric Telegraph" 63).

5. Carefully analyzing Galvani's and Alessandro Volta's experimental strategies, Marcello Pera challenges the scientists' own claims that their hypotheses were

based on "facts and ratiocination." Pera argues that their disagreement about the existence of animal electricity was more a struggle of "hidden metaphysics," "a clash of assumptions or interpretive theories that functioned as gestalten" (xxii, xxv).

6. In 1937, in a speech celebrating the second centenary of Galvani's birth, Q. Maiorana had no qualms about comparing organic and technological communications systems. The phenomena that Galvani had observed in frogs, he proposed, "contain[ed] the germ of modern wireless telegraphy. . . . What in Galvani's hands could move a muscle, brought Marconi's voice across oceans" (Galvani xi).

7. In a recent study, Joost Mertens provides strong evidence against the traditional view that Volta constructed the battery as the definitive evidence against Galvani. Instead, Mertens proposes, Volta already believed he had refuted Galvani's results, and his real goal was to build a public demonstration device. Volta had consciously pursued such a strategy for decades, confident that the success of technological devices based on his theories affirmed those theories' truth (Pera 45–46).

8. Over the centuries, scientists' descriptions of the mechanism of nerve-impulse conduction have closely paralleled theories about waves and particles in physics. Johannes Müller wrote in the 1830s that "every cause which produces a sudden change in the relation of [a nerve's] molecules to each other . . . excites a muscular contraction" (1: 688). Hermann von Helmholtz, who performed key experiments in both thermodynamics and electrophysiology, called heat "a peculiar shivering motion of the ultimate particles of bodies" (*Science and Culture* 28). In 1885, as physicists grew increasingly interested in waves, George Romanes described the nerve impulse as "an invisible or molecular wave of stimulation" (25). Whatever the nervous principle was, scientists believed, it must be closely related to energy itself, and their descriptions of it reflected their understanding of energy's interrelated forms.

9. Timothy Lenoir argues that DuBois-Reymond's increasingly sensitive instruments shaped not only the mechanisms he proposed to explain the way nerves worked but the questions he asked about the nervous system. Lenoir calls him "the Faraday of physiology" ("Models and Instruments" 4–7).

10. Explaining the appropriateness of his use of the term "electrotonic" for the excited state of a nerve, DuBois-Reymond wrote that the word "recalls to mind the striking analogy . . . between the law of the galvanic contractions and the law which regulates the induction of currents in a circuit by moving a magnet. . . . Faraday supposed that in a conductor in which a current was induced a change took place which lasted as long as the inducing current itself continued. . . . Faraday called this altered condition . . . the electrotonic state" (*On Animal Electricity* 185–86).

11. According to Kathryn M. Olesko and Frederic L. Holmes, when DuBois-Reymond translated Helmholtz's article into French so that Alexander von Humboldt could present it to the French Academy of Sciences, he recommended substantial organizational changes because he found the original draft "obscure" and likely to produce misunderstandings.

12. It was DuBois-Reymond who favored opening with references to a time interval "not too difficult to estimate" (Olesko and Holmes 91–92). The original

German reads: "Ich habe gefunden, dass eine messbare Zeit vergeht, während sich der Reiz, . . . bis zum Eintritt des Schenkelnerven in den Wadenmuskel fortpflanzt" ("Vorläufiger Bericht" 71). My translation.

13. Like Helmholtz, DuBois-Reymond was following the precedent of astronomy, where it had long been known that the "real time" of events could not be equated with observers' divergent recordings. In 1864 the Swiss astronomer Hirsch, who improved Helmholtz's apparatus and repeated his experiments, had proposed a system by which two astronomers might determine the "real time" of an event by comparing their "personal corrections." As a counterpart to "real time," Hirsch proposed the term "physiological time" (DuBois-Reymond, "On the Time Required" 130, 127).

14. While Babbage never formally studied the nervous system, it is possible that he learned about physiology from Alexander von Humboldt. When Babbage visited Berlin in the late 1820s, Humboldt received him enthusiastically and gave him a list of all of the "savants" who were currently in the city, offering to introduce him to any that he desired to meet (Moseley 95).

15. By creating an engine that performed calculations, Babbage realized, he was performing a translation. The motions of its parts were a language, "an expression of the accretion of quantity," so that every calculation converted variables and operations into movements and then back into tables of fixed digits (Buxton 46).

16. Babbage's biographer Anthony Hyman believes that this paraphrase of Buxton's accurately represents Babbage's position and reflects Buxton's and Babbage's extensive discussions of the calculating engines (Hyman xiv).

17. Helmholtz's relationship with Kantian epistemology is more complex than it might first appear. While Timothy Lenoir argues for a "final break" with Kant (and Müller) by 1867, S. P. Fullinwider believes that Helmholtz "consider[ed] himself a true Kantian up to the end of his life," and P. M. Heimann finds that even the third volume of Helmholtz's *Optics* (1867) contains "characteristically Kantian phraseology" (Lenoir, "Eye as Mathematician" 153; Fullinwider 41; Heimann 221). Fullinwider agrees with Lenoir, however, that Helmholtz "transformed Kantian epistemology" by proving that much of what Kant considered to be intuition was actually learned empirically (Fullinwider 41, 45).

18. For a systematic study of Helmholtz's use of the term "*Zeichen,*" see Dosch, "Concept of Sign and Symbol." For a discussion of Müller's and Helmholtz's different uses of the term "sign," see Lenoir, "Eye as Mathematician" 110–18.

19. The original German reads: "Insofern die Qualität unserer Empfindungen uns von der Eigenthümlichkeit der äusseren Einwirkung, durch welche sie erregt ist, eine Nachricht giebt, kann sie als ein Zeichen derselben gelten, aber nicht als ein Abbild. Denn von einem Bilde verlangt man irgend eine Art der Gleichheit mit dem abgebildeten Gegenstande. . . . Ein Zeichen aber braucht gar keine Art der Aehnlichkeit mit dem zu haben, dessen Zeichen es ist" (Dosch 49).

20. The original German reads: "Das 'Ding an sich' (das würde eben die reine folgenlose Wahrheit sein) ist auch dem Sprachbildner ganz unfasslich und ganz und gar nicht erstrebenswerth. Er bezeichnet nur die Relationen der Dinge zu den Men-

schen und nimmt zu deren Ausdrucke die kühnsten Metaphern zu Hülfe. Ein Nervenreiz zuerst übertragen in ein Bild! erste Metapher. Das Bild wieder nachgeformt in einem Laut! Zweite Metapher" (Nietzsche 879).

21. In a careful study of Nietzsche's scientific reading between 1872 and 1875, Karl Schlechta and Anni Anders demonstrate Nietzsche's ongoing interest in Helmholtz's work. Nietzsche is known to have read Helmholtz's essay "On the Interaction of Natural Forces" (1854) and in April 1873 to have borrowed (and probably read) his *Handbook on Physiological Optics* (1856–67). In 1868, Nietzsche intended to study Helmholtz's writings systematically but seems never to have fulfilled that goal. Instead, he learned about Helmholtz's ideas mainly through other writers such as F. A. Lange and J. C. F. Zöllner (Schlechta and Anders 67–68, 123–25). I am grateful to Geoff Waite for bringing Schlechta and Anders's study to my attention.

22. Ernst Florens Friedrich von Chladni (1756–1827), who founded the modern science of acoustics, devised an ingenious system for representing sound waves visually. He coated metal plates with fine powder, which moved to the nodal points when intense sounds caused the plates to resonate. Thanks to Chladni's experiments, it became possible to "see" sound. I am indebted to Sander Gilman for informing me about Chladni's work. Schelchta and Anders paraphrase Nietzsche's presentation of this metaphor as follows: "Die '*darunterliegende*' Nerventätigkeit, selbst durch ungestaltete Empfindungs-Reize verursacht, projiziert in dieser Fläche Formen, die ihrerseits wieder '*allerzarteste Lust- und Unlustempfindungen*' hervorrufen: die Empfindung des Bildes. Wobei die gleiche Nerventätigkeit immer wieder das gleiche Bild erzeugt. Das Bild selbst verhält sich zu der '*darunterliegenden*' Nerventätigkeit wieder '*wie die Chladnischen Klangfiguren zu dem Tone selbst*'" (Schlechta and Anders 108, original emphasis.)

23. Evaluating Helmholtz's view of nerve impulses and sensations as signs, Dosch writes that Helmholtz "was led to his concept of signs which saw the immense separation between the sensations and the outside world" but qualifies that "he also laid emphasis on the lawful relations between the signs" (54).

Chapter 2

1. The original German reads: "Sehen Sie nun wohl die Seele im Gehirn, als der einzig empfindlich bewussten Region des Körpers sitzen, und den ganzen übrigen Körper wie eine todte Maschine in ihrer Hand? So pulsiert in dem sonst bis zur Verödung centralisierten Frankreich nur in Paris das Leben der grossen Nation. Aber Frankreich ist nicht der richtige Vergleichspunkt, Frankreich wartet noch auf einen Werner Siemens, um es mit einem Telegraphennetz zu überspinnen. Denn wie die Centralstation der elektrischen Telegraphen im Postgebäude in der Königsstrasse durch das riesenhafte Spinngewebe ihrer Kupferdrähte mit den äussersten Grenzen der Monarchie im Verkehr steht, so empfängt auch die Seele in ihrem Bureau, dem Gehirn, durch ihre Telegraphendrähte, die Nerven, unaufhörlich Depeschen von allen Grenzen ihres Reiches, des Körpers, und theilt nach allen Richtungen Befehle an ihre Beamten, die Muskeln, aus" (*Reden* 2: 50–51). My translation.

2. Diderot's "characters" in *Le Rêve de D'Alembert* are based on real people: mathematician Jean Le Rond D'Alembert (1717–83), physician Théophile de Bordeu (1722–76), and Julie de l'Espinasse, all close friends of the author. When D'Alembert and de l'Espinasse read the dialogue, they found it so "personal" that they asked Diderot to burn it. Fortunately, he kept a copy (Fellows 101).

3. All translations of Diderot's works quoted here are my own. Diderot offers the crayfish metaphor as follows: "Donc tout le système nerveux consiste dans la substance medullaire du cerveau, du cervelet, de la moëlle allongée, et dans les prolongements de cette même substance distribuée à differentes parties du corps. C'est une ecrevisse, dont les nerfs sont les pattes, et qui est diversement affectée selon les pattes" (*Eléments de Physiologie* 87).

4. The original French reads: "Dans l'homme, les nerfs viennent se réunir et se perdre dans le cerveau; ce viscère est le vrai siège du sentiment; celui-ci, de même que l'araignée que nous voyons suspendue au centre de sa toile, est promptement averti de tous les changemens marqués qui surviennent aux corps, jusqu'aux extrémités duquel il envoie ses filets ou rameaux" (125).

5. Describing the rationale behind the physical arrangements of early nineteenth-century institutions, Michel Foucault explains: "Disciplinary space tends to be divided into as many sections as there are bodies or elements to be distributed. . . . Its aim was to establish presences and absences, to know where and how to locate individuals, to set up useful communications, to interrupt others, to be able at each moment to supervise the conduct of each individual, to assess it, to judge it, to calculate its qualities or merits. . . . the disciplinary space is always, basically, cellular" (143).

6. Unless otherwise indicated, all translations of Golgi's French and Italian articles are my own. I include the original only for quotations of substantial length in which the anatomist's choice of words is highly significant. Otherwise, key words that are not easily rendered into English will be given in parentheses in the main text. Here the translation is Clarke's and O'Malley's, and the French original reads: "Il est probable que ces innombrables subdivisions s'anastomosent entr'elles, et forment ainsi un véritable plexus, mais l'extrême complication de ce réseau ne me permet pas d'affirmer quelque chose de positif à ce sujet."

7. The original French reads: "Si les fibres nerveuses ne dérivent ni directement, ni indirectement des prolongements protoplasmiques, s'il n'y a de communication entre les différents groupes de cellules du système nerveux, ni par les anastomoses, ni par le réticulum diffus, quel est alors le mode d'origine de la fibre nerveuse dans la substance grise? Comment s'établit entre les diverses cellules ou les diverses régions du système nerveux cette relation fonctionelle que l'on est forcé d'admettre à priori?" Clarke's and O'Malley's translation.

8. The original Italian reads: "Dal momento che gli studi sull' elettricità dimostrano che le correnti elettriche possono effetuarsi senza continuità diretta delle parti conduttrici . . . perchè non potrebbesi ammettere che identiche leggi valgano anche pel sistema nervoso?" ("La rete nervosa" 593).

9. The original French reads: "Il n'était plus nécessaire d'invoquer une connexion matérielle, une fusion entre une fibre et l'autre, pour se rendre compte des rap-

ports fonctionels qui courent entre les différents groupes de cellules et entre les différentes parties du système nerveux centrale. . . . il n'y avait pas lieu de croire que la continuité directe entre les fibrilles de différente provenance était une condition *sine quâ non* pour la transmission de l'excitation des unes aux autres" ("La doctrine du neurone" 15–16).

10. The original French reads: "dont le prolongement nerveux se subdivise d'une manière compliquée, perd son individualité et prend part *in toto* à la formation d'un réticulum nerveux qui traverse toutes les couches de la substance grise" ("Recherches sur l'histologie" 298). Interestingly, references to "losing one's individuality" are preserved as a formula in all discussions of nerve cells in the 1880s and 1890s, whether the language is Golgi's original Italian ("perde la propria individualità") ("Sulla fina anatomia" 322); his original French ("perd son individualité") ("Recherches sur l'histologie" 298); Spanish ("perder su individualidad") (Ramón y Cajal, "Conexión general" 486); or German ("seine Individualität aufzugeben") (Waldeyer 1215). Whether scientists supported or opposed the idea of a nerve net, they shared the premise that to merge physically with another structure was to give up one's individual identity.

11. Michael Hagner argues that Golgi's intense opposition to cerebral localizationism was the most important factor preventing him from seeing individual nerve cells under the microscope (Hagner, "Les Frères ennemis," 41). While the drive to map the brain inherited some force from early nineteenth-century phrenological studies, it acquired fresh energy in 1861 when Paul Broca (1824–80) associated motor aphasia (the inability to speak) with the left temporal cortex. Broca's study was followed by Carl Wernicke's (1848–1905) identification of a center for language comprehension and by Gustav Theodor Fritsch's (1838–1907), Edouard Hitzig's (1838–1927), and David Ferrier's (1843–1928) associations of movements in specific body parts with areas of the motor cortex. Golgi was thus opposing a strong, successful field of neurology when he spoke out against brain localization as late as 1906. While most of Broca's, Wernicke's, Frisch's, Hitzig's, and Ferrier's associations of structure with function are still accepted today, most twentieth-century neurologists agree with Golgi that the nervous system works as a whole and that one must exercise extreme caution in associating a particular activity with a given area of the brain. For an insightful analysis of nineteenth-century brain localization debates, see Michael Hagner's *Homo cerebralis*.

12. The original French reads: "Dans la plupart des centres nerveux ce n'est point du tout le rapport isolé et individuel entre les fibres et les cellules qui a lieu, mais au contraire une disposition évidemment destinée à permettre la plus grande varieté et la plus grande complication dans leurs rapports mutuels. . . . *la disposition fondamentale des éléments centraux indique une tendance à effectuer les communications les plus étendues et les plus compliquées et non des rapports restreints et isolés*" ("Recherches sur l'histologie" 303, original emphasis).

13. Ramón y Cajal achieved this result by improving Golgi's technique, repeating the staining process so that the tissue was more thoroughly penetrated, and by working on preparations from young or embryonal animals. After he presented his

findings to the German Anatomical Society in Berlin in 1889, the leading neu-roanatomists—most of whom were present and could see the freely terminating neu-rons with their own eyes—conceded that nerve cells did not merge into a net (Hag-ner, "Les Frères ennemis," 43).

14. All translations of Santiago Ramón y Cajal's Spanish and French publica-tions are my own.

15. The original Spanish reads: "Champolion, adivinando el lenguaje muerto de los geroglíficos egipcios, y Layard y Rawlinson, desentrañando el misterioso sentido de los caracteres cuneiformes de las inscripciones de Ninive y Babilonia, se han prop-uesto problemas mucho más sencillos que los neurólogos" (*Textura del sistema nervioso* 18). After Thomas Young's attempt to decipher the Egyptian hieroglyphics on the Rosetta Stone, Jean François Champollion (1790–1832) demonstrated the relation of these characters to the stone's Greek inscription in 1821. Decoding the hieroglyphics proved challenging because some of the signs were alphabetic, whereas others represented syllables or even entire ideas. Sir Austen Henry Layard (1817–94) attempted to locate and excavate the ancient cities of Mesopotamia (now Iraq) and discovered cuneiform tablets that revealed a great deal about Babylonian culture. His *Discoveries in Ruins of Nineveh and Babylon* (1853) was widely read in Europe (*New Encyclopedia Britannica* 3: 75). It is interesting that as an investigator of the nervous system, Ramón y Cajal compares himself both to an archaeologist and to a linguist. As he studies the relations between neurons, he implies, he is both the explorer seeking sources of information and the close reader struggling to interpret them.

16. The original Spanish reads: "¿Tocan realmente el protoplasma desnudo de la célula o existe entre ambos factores de la sinapsis membranas limitantes?" ("¿Neu-ronismo o reticularismo?" 229).

17. The original Spanish reads: "la conexión entre éstas y los cilindros ejes puede ser mediata y verificarse la transmisión de la acción nerviosa como las corrientes eléctricas de los hilos inductores sobre los inducidos" ("Estructura de los centros nerviosos" 315).

18. The original Spanish reads: "El hecho mismo de la transmisión de la onda desde una neurona à otra, obedecería a fenómenos químicos: en realidad, el impulso provoca un cambio químico en las arborizaciones nerviosas, el cual, obrando à su vez como estímulo fisico-químico sobre el protoplasma de otras neuronas, crearía en éstas nuevas corrientes. El estado consciente estaría precisamente ligado a los cam-bios químicos suscitados en las neuronas por las terminaciones nerviosas" (*Textura del sistema nervioso* 1149).

19. The original French reads: "La cellule nerveuse présente un appareil de *réception* des courants figuré par les expansions dendritiques et le corps cellulaire, un appareil de *transmission* représenté par le prolonguement cylindraxile, et un appareil de *répartition* ou de *distribution* représenté par l'arborisation nerveuse terminale" (457, original emphasis).

20. The original French reads: "Vis à vis de la théorie des réseaux celle des arborisations libres des expansions cellulaires susceptibles de s'accroître apparaît

non seulement comme plus probable, mais aussi comme plus encourageante. Un réseau continu pré-établi—sorte de grillage de fils télégraphiques où ne peuvent se créer ni de nouvelles stations ni de nouvelles lignes—est quelque chose de rigide, d'immuable, d'immodifiable, qui heurte le sentiment que nous avons tous que l'organe de la pensée est, dans certaines limites, malléable et susceptible de perfection, surtout durant l'époque de son développement, au moyen d'une gymnastique mentale bien dirigée. Si nous ne craignions pas d'abuser des comparaisons, nous défendrions notre conception en disant que l'écorce cérébrale est pareille à un jardin peuplé d'arbres innombrables, les cellules pyramidales, que, grâce à une culture intelligente, peuvent multiplier leurs branches, enfoncer plus loin leurs racines, et produire des fleurs et des fruits chaque fois plus variés et exquis" (467–68).

21. In his study of Lewes's defense of vivisection, Richard Menke reports that Lewes had always wanted to be a physician but left medicine because he could not bear to see patients in pain ("Fiction as Vivisection" 619).

22. While the rapid development of telegraphic networks accounts for Helmholtz's and DuBois-Reymond's use of the telegraph metaphor in the 1850s, it does not explain why scientists continued to use it in the 1860s and 1870s when Lewes, by 1877, found it worthless. Possibly their reliance on telegraphy early in their careers to create models of neuronal transmission led them to retain the metaphor even when telegraphic networks were changing from an exciting novelty into a dreary cultural fact.

Chapter 3

1. Patrick J. McCarthy proposes that *Middlemarch*'s webs were "derived from the scientific vocabulary" of Lewes and Xavier Bichat (805).

2. P. di Pasquale finds that "Eliot's ideas are often . . . embodied in images," and Selma Brody believes that Eliot uses metaphors as scientists use their models (P. di Pasquale 426; Brody, "Physics in *Middlemarch*" 52).

3. In Europe, the standardization of time occurred surprisingly late, and it was the growth of society's new nerve networks that brought it about. Henry Booth's 1845 pamphlet urging the British to adopt standard time depicts a centralized communications network as a national nervous system acting as a will in a national body. "There is sublimity in the idea of a whole nation stirred by one impulse," he wrote, "in every arrangement one common signal regulating the movements of a mighty people" (Simmons 346). Before the early 1840s, it had been generally accepted that time varied from place to place, but "nobody moved fast enough to notice" (Simmons 345). The development of national railways made standardization essential, not just for the maintenance of schedules but for the avoidance of deadly accidents. In 1852, a special telegraph line was constructed to transmit time signals from the Royal Observatory along most of the major British railways, and Greenwich Mean Time was adopted throughout England (Simmons 346). While a few voices protested "railway-time aggression" as an encroachment on local authority, most cities accepted the standard. Brussels and Paris were connected to the Royal

Observatory in 1853 and 1854, respectively, and its signals were soon received throughout Europe (Simmons 346). The railways created a new consciousness of time, and their timetables encouraged people to plan not just their travels but many of their daily activities in terms of prearranged schedules (Simmons 184). By allowing people to predict the movements of bodies and machines, the timetable proved ideal for organizing the lives of students and workers. Although designed for the railways, it was "widely adaptable to the closer regulation of society at large . . . a fusion of the new language of technology with the familiar one of order and control" (Simmons 184).

4. In the United States, where land was cheaper than labor, the rails were laid according to geographical contours, and the technical requirements of these curved lines led to the development of longer, more open railway cars. The railway cutting, resulting from the demand for straight, level lines, was far more common in England than in America (Schivelbusch 96–99).

5. The conventions of nineteenth-century iconography suggest that Hudson is the spider, not the fly. A. E. Brehm's popular *Illustriertes Thierleben* (1869) depicts spiders in exactly the same position (684–85).

6. John Clark Pratt, who edited Eliot's *Middlemarch* notebooks, has been unable to locate the source of this reference, but 280,000 miles per second is close to Wheatstone's figure for the velocity of electricity. This figure was frequently cited in Victorian journals, as in the unsigned article "Electricity and the Electrical Telegraph" in *The Cornhill Magazine* (67).

7. From this point on, Eliot wrote steadily, completing the novel in September 1872. It was published in four installments between December 1871 and December 1872.

8. Joy Hooton observes that Eliot is "at pains to uncover the effects of Time" in *Middlemarch* and that she depicts it as a "dense, spatial medium" (188).

9. Di Pasquale argues that the plot of *Middlemarch* was "consciously based" on a web as a structural framework (428).

10. In an important new reading of Eliot's novel, Selma Brody proposes that *Middlemarch* contains "an ingenious model of society which is based on the Kinetic Theory of Gases" ("Physics in *Middlemarch*" 42).

11. Lewes was every bit as intrigued by Tyndall's particles as Eliot was, and an understanding of kinetics was as essential to his physiological writings as it was to her realistic fiction. In May 1871, Lewes was studying "the highest problems in statics and dynamics" (Haight 435). A reference in Eliot's notebook to "waves of molecular motion" may reflect her exchanges with either Tyndall or Lewes, but most importantly it shows the close affinity of their projects to her own (Pratt 48).

12. Patrick J. McCarthy believes that Eliot's study of cell theory "clearly played a part in her choice of central metaphor for the novel" (815). Histology, he points out, is derived from the Greek word "*histos*," or "web."

13. In his textbook *General Anatomy* (1801), French anatomist Xavier Bichat had argued that bodily organs were really compounds of twenty-one tissues. He proposed that physicians could understand how bodies worked, in health and disease,

only if they understood how these tissues interacted. Eliot and Lewes owned French editions of Bichat's *General Anatomy* (1801) and *Physiological Researches on Life and Death* (1800), so it is very likely that Eliot knew Bichat's arguments.

14. Much later in the novel, the narrator refers to "possible histories of creatures that converse compendiously with their antennae, and for aught we know may hold reformed parliaments" (541). The image mirrors Mary's earlier one and playfully suggests the pettiness of local politics.

15. J. Hillis Miller argues that in her fiction, Eliot used a new, more "realistic" kind of synecdoche. In *Middlemarch,* the part may be taken as a sampling of the whole but never as a true representation of it. Interestingly, the novel's three "totalizing metaphors"—the labyrinth, "flowing water," and "woven cloth"—are built on the assumption that "the structure or texture of small-scale pieces of the whole is the same as the structure or texture of the whole," suggesting that one can learn about the whole by investigating the sample. Miller believes that this contradiction in Eliot's metaphorical system destabilizes her text ("Optic and Semiotic" 126, 129).

16. In Helmholtz's influential essay "On the Conservation of Force" (1847), the physicist cited Morse's electric telegraph as one of the clearest examples of the connection between electricity and magnetism, two different but closely related forms of energy (*Science and Culture* 102).

17. According to John Clark Pratt, the greatest problem Eliot encountered when writing the early, "medical" version of *Middlemarch* was that of cause and effect. Pratt proposes that fevers appealed to her as a subject for Lydgate's research because they show the operation of cause and effect in the body.

18. I am indebted to Lawrence Rothfield and his students at the University of Chicago for their analysis of this metaphor.

19. These images suggest Maxwell's demon, which violates the second law of thermodynamics by selectively allowing certain particles to move through a small opening from a cooler to a warmer compartment. Proposed by William Thomson in 1852, this law states that whenever energy changes forms, some is "lost" and can never again be employed to do useful work. Maxwell proposed this thought experiment in 1871, while Eliot was writing her novel. One far-reaching consequence of Maxwell's statistical mechanics has been twentieth-century scientists' interest in the impact of small events on complex systems. When Eliot studies how seemingly trivial local events can have tremendous repercussions and how global changes can create unexpected local effects, she anticipates the thinking of twentieth-century dynamics.

20. Generally, Eliot presents Lydgate's scientific endeavors in a positive light, and this fragmentation of living organisms into bits does not fairly represent his approach to anatomy. Tess Cosslett believes that Lydgate's scientific research serves as the "central moral metaphor" of the novel because of its emphasis on careful observation and the adjustment of hypotheses to external realities (77). Both Eliot and Lewes were critical of experimental science's tendency to break complex systems down into bits, and while they supported its goals, they worried that experimentalists might forget the living wholes from which these bits were taken.

21. Gordon Haight writes that one possible model for Casaubon was Eliot's first employer, Dr. Brabant, a man "whose fastidiousness made his work something like Penelope's web" (Haight 50–51). Like most critics, however, Haight believes that Casaubon is Eliot's own creation and is not based on any actual person.

22. When discussing the development of the nervous system in *The Physical Basis of Mind,* Lewes wrote: "The metamorphoses do not take place by a gradual modification of the existing organs and tissues, but by a *resolution* of these into their elements, and a *reconstruction* of their elements into tissues and organs. The muscles and nerves . . . undergo what may be called a *fatty degeneration,* and pass thence into a mere blastema. It is out of these ruins of the old tissues that the new tissues are reconstructed" (292, original emphasis). This image of an organic struggle between "Old and Young" is another that he and Eliot shared.

23. François Raspail's (1794–1878) *Nouveau système de chimie organique* (1833) may have inspired Lydgate's quest for the origin of all tissues. After recovering an "amorphous insoluble mass of debris" from boiled animal tissue, Raspail proposed that he had discovered the fundamental organic substance (Forrester 3; Greenberg 45–46).

24. In *Middlemarch,* Eliot plays with all possible meanings of communication. When Bambridge leads Fred through "unsanitary Houndsley streets" in search of the ill-fated horse Diamond, Fred brings back not just a "bad bargain in horse-flesh" but the typhoid fever bacillus (179).

25. This body of gossip bears a certain resemblance to Tyndall's and Maxwell's gases. By calling it "free oral communication," Selma Brody shows its relation to the energy or collective properties of a large population of particles. Brody writes perceptively that "a great many voices are averaged to make one unit of gossip" ("Physics in *Middlemarch,*" 50). Much to Lydgate's despair, it is only in the network of gossip that he will "tell appreciably on the averages" (Eliot, *Middlemarch* 100).

26. Joy Hooton observes that in *Middlemarch,* "misconceptions and failures of communication are built into the nature of things" (200).

27. Gillian Beer writes that in *Middlemarch,* "there are gaps of judgment and feeling over which language will not seamlessly spread itself" (48). J. Hillis Miller proposes that Eliot's characters persistently misread the world's signs because of "the universal human habit of thinking by analogy" ("Roar on the Other Side" 237). Less optimistic than Eliot about the possibility of real knowledge, he believes that only chaos underlies these signs. According to Miller, the novel teaches one how to live not in the world but in the world of language.

28. J. Hillis Miller believes that this "wadding" is each individual's metaphorical system, which protects him or her from a direct view of the chaos underlying the world's signs ("Roar on the Other Side" 241).

29. While critics have proposed several "sources" for Eliot's pier-glass parable, none have yet suggested Faraday's lines of force. N. N. Feltes believes that Eliot took the image from Herbert Spencer, who was interested in Ruskin's "optical delusion" of a moonbeam that appears to follow one over a body of water (69). Since Spencer did not publish his thoughts on this "delusion" until 1873, it is more likely that he

learned of the image from Eliot. Selma Brody's proposal that the metaphor has its roots in Tyndall's optical experiments is far more reasonable ("'Pier-Glass' Image" 57). The real source, of course, is Eliot's own creative imagination, as fertile in thought-inspiring fictions as those of the scientists with whom she interacted.

30. Sally Shuttleworth finds that Eliot analyzes action as a "play of force." According to Shuttleworth, "the self in *Middlemarch* is not a predefined entity that determines action" but a "product of convergence of forces" (159–60). For Eliot, the individual is a cross-point in the social web.

31. Perhaps because of desire's close relation to the will, and hence to the nervous impulse, Eliot describes her characters' erotic feelings with the language of electricity and magnetism. If desire weaves webs, it is also a force transmitted through the webs of one's nerves. Lydgate's galvanic experiments reflect his intense attraction to Laure, and when the doctor later desires Rosamond, we hear of her "torpedo contact" (*Middlemarch* 456; Greenberg 48). Both women shock Lydgate, and he spends the latter part of his life stunned into immobility by his wife's materialistic desires. Most often, Eliot's references to electrical forces describe the more complex attraction between Will and Dorothea. The narrator reports that "when Mrs. Casaubon was announced, [Will] started up as from an electric shock, and felt a tingling at his finger-ends. Any one observing him would have seen a change in his complexion, in the adjustment of his facial muscles, in the vividness of his glance, which might have made them imagine that every molecule in his body had passed the message of a magic touch. . . . Will . . . was made of very impressible stuff" (268). While Lydgate's "shocks" present electricity—and desire—as a violent force, reminding the reader of its power to destroy life, Eliot's use of it to describe Will and Dorothea's feelings emphasizes the nervous system's acute sensibility, its capacity to be influenced by outside forces. Will's and Dorothea's electric shocks—and there are many—chart not just their developing attraction but their development as moral beings.

Chapter 4

1. This passage is taken from a letter of 15 February, 1838, to Morse's collaborator F. O. J. Smith.

2. This passage is from the poem "By Telegraph" in the anthology *Lightning Flashes and Electric Dashes,* edited by William John Johnston.

3. Jeffrey Kieve and George Sauer both believe that "C.M." was Charles Marshall of Paisley, who may have withheld his identity for fear of being accused of practicing magic (Kieve 14; Sauer 5).

4. Writing to his mother about the telegraph, Cooke declared on 5 April, 1836, "I prepared to make a model" (5). By November, when he grew discouraged, he wrote to her that "my mind is nearly made up to return to modelling if the instrument does not answer . . . I shall make my preparation for renewing my labors in wax" (12).

5. Dr. Charles Jackson of Boston, who told Morse about these experiments, later claimed that he had given Morse the idea of using electricity to transmit information (Morse 2: 6; Briggs and Maverick 25–26; Prescott 58).

6. In 1837, the U.S. Congress had invited scientists and engineers to submit designs for a national telegraph line that would stretch from New York to New Orleans. They had a semaphore system in mind, for at that time "telegraph" referred to visual communications systems. Morse's model was the only one submitted that relied on electricity, and many congressmen were skeptical about the newfangled system he was proposing (Standage 42–47).

7. In the United States, telegraph lines sometimes preceded those of the railways. A transcontinental telegraph was operational eight years before the transcontinental railroad (Carey 203).

8. Tom Standage makes the important point that even after the first telegraph lines were constructed, it took several years before telegraphy was accepted by the public as a convenient, reliable form of communication. In the United States, the telegraph was initially regarded as a financial failure (Standage 48–55).

9. Originally, Morse had envisioned a system of secret cipher in which numbers recorded in carefully guarded telegraphic dictionaries would stand in for words and phrases. The dots and dashes transmitted would represent these numbers, not the words directly. This system proved so inefficient, however, that Morse shifted to a more accessible one in which dots and dashes corresponded directly to letters of the alphabet.

10. Katie Louise Thomas argues that British nineteenth-century fiction "reveals a cultural construction of the telegraph worker as hyper-public and sexualized." American stories suggest a similar image.

11. Intertwined with late nineteenth- and early twentieth-century Taylorist discourse that "model[ed] human movements after machines," Katherine Stubbs has detected "a somewhat different discourse . . . which articulated a correspondence between machines and certain types of human bodies" (264). This "rhetorical construction" presented women as more machinelike than men, more adept at involuntary movements, and hence better suited for working with machines. I am grateful to N. Katherine Hayles for bringing Stubbs's work to my attention.

12. In a study of telegraphic style, Kay Yandell describes operators' exhilarating experiences of freedom in their on-line exchanges. For female operators, she proposes, the telegraph "liberate[d] the speaker's voice from all gender markers" and "allow[ed] woman to speak for the first time on an equal footing with man."

13. In her discussion of Ella Cheever Thayer's *Wired Love*, Ellen McCallum argues that the unstable identities constructed through a telegraphic relationship reveal cultural anxieties about changing gender roles in the late nineteenth century. The fact that women could support themselves and lead independent lives as telegraphers threatened the traditional image of men as breadwinners. Churchill's story, in which both men and women pass for the opposite sex over the wire, exposes these cultural concerns to an even greater degree than Thayer's novel.

Chapter 5

1. In *Bodies and Machines,* Mark Seltzer points out that "the turn-of-the-century fascination with technologies of writing and representation inheres not simply

in the notion that machines *replace* bodies and persons . . . nor is it accounted for primarily in the notion that persons are *already* machines . . . nor even is it "covered" in the notion that technologies *make* bodies and persons. . . . What makes it possible for these powerfully insistent, but not entirely compatible, notions to communicate on another level is the radical and intimate *coupling* of bodies and machines" (12–13).

2. In "Love Lines, Crossed Wires," Ellen McCallum identifies the novel's "secondary aim" as "seducing the audience to like technology." I am grateful to McCallum for introducing me to Thayer's novel. Kay Yandell also proposes that a great deal of telegraphic literature tried to explain the details of telegraphy to the "uninitiated."

3. As Ellen McCallum points out, "crossed wires have been a staple of romance plots long before the invention of the telegraph," but "telegraphy particularly transforms the mistaken identity and mistaken assumptions that fuel romance plots." By introducing a new technology of communication to a traditional literary form, Thayer raises questions about how identities are socially constructed.

4. Nattie's feelings about her job contradict Tom Standage's finding that telegraphy was a meritocracy in which people could advance regardless of class background. Standage cites the examples of Andrew Carnegie and Thomas Edison, both of whom began their careers as telegraph boys. "I do not know a situation in which a boy is more apt to attract attention, which is all a really clever boy requires in order to rise," Carnegie later reflected (Standage 64). The difference among Standage's, Thayer's, and Henry James's depictions of telegraphers is probably due to gender. Young female telegraphers were not expected to advance, only to resign when they married and be replaced by other young women, receiving no pensions or benefits (Moody 56).

5. Noses and teeth, whose function and appearance suggest a telegraph key, constitute a running motif in Thayer's novel. Usually they are associated with repulsive characters. Celeste Fishblate has an uncommonly large nose and teeth, and the impostor posing as "C" has "obtrusive, fighting teeth" (102). Proposing to Celeste (who he thinks is Nattie), Quimby cries, "I fell in love with you ever since I saw your nose . . . I fell in love with your nose!" When Cyn asks Nattie straight out whether she thinks Nattie and Clem are in love, the narrator tells us, "had Cyn drawn forth a Bowie knife and clipped off her nose, she could not have been more astounded" (205, 237). Representing the body and its primitive drives, noses and teeth often assume a threatening aspect in *Wired Love*.

6. Paul Eros writes that in telegraphic communication, "the sign itself becomes increasingly complex as its signified is coinstantaneously eroded. Telegraphy widens the chasm between the artifice of language and its motivation in the human mind."

7. In 1853, the Electric Telegraph Company of Great Britain charged one shilling to send a telegram within a radius of fifty miles, two shillings sixpence to send one between fifty and a hundred miles, and five shillings to send one over a hundred miles. At the time, these prices would have been too expensive for most individuals but within the reach of prospering businesses (The Electric Telegraph Company Chart 1853).

8. Placing the protagonist's frustration in sociohistorical context, Dale M. Bauer and Andrew Lakritz write that "the telegraphist represents the convergence or tantalizing proximity of classes within the context of an emerging service economy" (61). Her endless counting of words shows how Taylor's vision of efficient, machine-like movements was affecting workers' lives. When James's character compares her work to sitting in the stocks, she reveals the humiliation of such service-oriented positions (Bauer and Lakritz 63).

9. I am grateful to Laura Demanski for her helpful comments on my analysis of "In the Cage." I am especially indebted to Demanski for her point that the tenuous, problematic relations depicted in James's story of telegraphy recur throughout his major works.

10. In Wright's liberal-humanist reading of the story, James's purpose is to show that all knowledge consists of fragments imaginatively connected. "In [the telegraphist's] plight," he writes, "we recognize the universal problem which besets us, too" (174).

11. Calling telegraphy and realism "chronologically twinned technologies," Richard Menke believes telegraphy suggests the strategies of nineteenth-century fiction ("Telegraphic Realism" 6). Like the realist novel, he proposes, telegraphy "confirms that unseen connections are already in place" and tries to interconnect separate spaces through discourse ("Telegraphic Realism" 4). In 1850, both telegraphy and realist fiction promised to "bind the world together with instant communication" ("Telegraphic Realism" 26). When James wrote "In the Cage," argues Menke, he questioned the binding claims of both.

12. People's uncertainty about the kind of knowledge telegraphy offered resulted partly from English's conflation of different ways of knowing into a single word. Romance languages distinguish personal acquaintance from the acquisition of information, as in the French *connaître* and *savoir*. German, too, has separate verbs for developing an acquaintance (*kennen*) and acquiring factual knowledge (*wissen*). Well aware of this deficiency, James indicates in his opening paragraph that telegraphy offers no *connaissance*. Although it promises to deliver valuable facts, its decontextualized statements offer only certain kinds of "knowledge."

13. Richard Menke proposes that the narrator's position suggests that of the realist narrator. The telegrapher would like to believe she can see into everyone she is connected to, but unfortunately she cannot. By "plac[ing] narrative language in the cage," James questions the strategies of realist fiction ("Telegraphic Realism" 15).

14. Jennifer Wicke writes that "the cage animalizes by encapsulating a perpetual class rage" ("Henry James's Second Wave" 146).

15. The high season for telegraphy was July through September, the low one December through February. One of the principal grievances of telegraphers was that the postal service forced them to take their vacations in the winter, granting them two weeks in the high season only after decades of service. By having his protagonist take her vacation in September, James reveals his lack of acquaintance with telegraphers' working conditions. In writing the story, he relied only on general

knowledge, more concerned with the epistemological aspects of the tale than with specifics about daily life (W. Stone 245–46).

16. Stuart Hutchinson reduces the protagonist's desire for power and autonomy to a fear of losing her virginity. Echoing Freud's account of female sexual development, he writes that "for too long, she has been taking the benefits of masculine stimulation without the compromise of reciprocation" (24). What she fears, however, is probably not sexual intercourse but the entrapment of motherhood and the life-threatening process of childbirth.

17. In interesting analyses, two critics compare James's protagonist to a spinner or weaver. Jennifer Wicke calls her an "Ovidian heroine at her shuttle" weaving a "class tapestry" ("Henry James's Second Wave" 147). Janet Gabler-Hover likens her to two of the three sisters of fate: Lachesis, who spins out the thread of one's life; and Atropos, who cuts it (Gabler-Hover 259). Both comparisons emphasize the girl's desire to control her customers' lives by controlling the threads of their communications.

18. In readings extremely hostile to the protagonist, several critics have condemned her desire for power. E. Duncan Aswell calls her "naive, spiteful, and continually engaged in self-deception," and Joel Salzberg finds her "shrewish," with a "ruling passion for superiority" (Aswell 375; Salzberg 68, 67) To Salzberg, James's story is an "almost clinical portrait of a girl dominated by narcissism and hostility" (66). Stuart Hutchinson calls Aswell's reading "cheaply sarcastic" but later reproves the protagonist for "taking the benefits of masculine stimulation" (19, 24). To these critics, who barely consider the protagonist's socioeconomic situation, her anger and quest to control others seem reprehensible. When one imagines her "position," however, such feelings in an intelligent and nearly powerless person become quite understandable.

19. Regardless of their theoretical perspectives, critics agree that "In the Cage" ends with a humiliating resignation. Ralph Norrman, who reads the protagonist's struggle as an attempt to "square" objective and subjective reality, argues that in the end, "the supremacy of objective reality is asserted" (425, 427). Dale M. Bauer and Andrew Lakritz, who read the story from a sociohistorical perspective, feel that the novella is about "coming to recognition of human limitation and restriction"—especially, I would argue, in an epistemological sense (62).

20. Dale M. Bauer, Andrew Lakritz, and Jennifer Wicke all argue that to some degree, the cage represents language itself. But it would be too simple, they qualify, to claim that "language is a prisonhouse from which the telegraphist must attempt to escape" (Bauer and Lakritz 67). Tying the cage to language, class, and sexuality, Bauer and Lakritz claim that "social structures do not merely inhibit or cage human potentiality . . . they produce eruptions of force" (67). As Jennifer Wicke puts it, "the cage is made *of* words, but not by them" ("Henry James's Second Wave" 151).

21. In a psychoanalytic reading of James's story, William Veeder argues that the protagonist's unconscious contains a repressed, internalized "toxic mother." I disagree with Veeder, however, that her anger is rooted primarily in childhood conflicts. Her comments about social inequities point to ongoing frustrations.

22. An article written for *The Cornhill Magazine* in 1860 suggests the protagonist is only partly correct in presuming most customers think she is stupid. At least some clients knew that operators came from different class backgrounds and expected their competence to vary accordingly:

> The rate of transmission varies greatly, being dependent not merely on the experience of the telegraphist, but on his education and quickness of comprehension. An intelligent operator would find no difficulty in reading forty words per minute, whilst an illiterate railway signalman would find *two* sufficient for *his* comprehension in an equal space of time. ("Electricity and the Electrical Telegraph" 70, original emphasis).

Of course, in the thirty-eight years that elapsed between the publication of this article and James's story, upper-class clients might have developed an equal contempt for all operators. In 1881, a writer for *Blackwood's Magazine* advised readers to "write every word [of a telegram] with the distinctness, not that you would consider sufficient in a letter to a friend, but that you would aim at in writing to an illiterate person" ("Freaks of the Telegraph" 478).

23. Leon Edel compares James's telegrapher to a detective, writing that she "practices, in a modest and simplified form, the deductive methods celebrated in Sherlock Holmes. . . . She is a detective of her own confined soul" (476). Holmes is by far the better sleuth.

24. Ralf Norrman calls the word "know" a "Jamesian ambiguity signal" and points out that James "harps on the word 'knowledge'" throughout the story (426, 427).

25. Critics disagree about the protagonist's ability to distinguish imagination from reality. E. Duncan Aswell proposes that she convinces herself they are separate, whereas in reality, they "crisscross at every point" (376). Stuart Hutchinson also believes that she sees a distinction—and maintains that there is one to see. Although she "wants the thrill of letting them touch," he writes, she consciously tries to keep them apart (20). Dale M. Bauer and Andrew Lakritz, however, find that she "confuses the intimacy of her knowledge gained as a telegraphist with the intimacy of persons" (63). While I find Aswell's and Hutchinson's interpretations more convincing in this case, it is clear that she wants reality to match her fantasies, so that at times she may let the boundary waver.

26. In a painstaking analysis of the story's plot, Ralf Norrman concludes that James used a minimalist numerical code for the lovers' communications because he was primarily interested in "social drama," not in the message or code itself. His primary goal, Norrman believes, was to "explore the various aspects of knowledge" (427). Dale M. Bauer and Andrew Lakritz agree, observing that the story centers around an "inscrutable text" that "seems to have both a central significance and an utter irrelevance" (61).

27. Janet Gabler-Hover sees gold as a "metaphor for [the protagonist's] emotions," one that the telegrapher employs herself (269).

28. Although the telegrapher in James's story has none of Dorothea's altruism,

both women long for recognition, the ability to make a meaningful difference in other people's lives.

29. In a letter to his brother William on 20 April, 1898, James wrote that he wished his sister Alice could have heard of William's election to the French Institut, "such a happy little nip as it would have given her" (James and James 352). Here, as in the story of the female telegrapher, those who "nip" are those who are excluded from power and privilege but spend their lives hearing of the doings of others.

30. In *Le Parasite*, Michel Serres argues that parasitism is never direct, occurring only where there is a relation between two other entities. In both biological and technological systems, the parasite is the "third term" that profits from and transforms this relation.

Chapter 6

1. Barrett hopes "to avoid any possibility of any information being gained by the ordinary channels of sense" or "derived through the ordinary channels of sense" (Barrett 75, 79). While radio and television did not exist at the time, twentieth-century writers interested in psychical phenomena continued to use Barrett's phrasing and compared thought transmission to their new technologies. In 1930, Upton Sinclair published *Mental Radio*, a study of his wife's purported thought reading abilities. In his introduction, Sinclair asks readers, "Can one human mind communicate with another human mind, except by the sense channels ordinarily known and used?" (3).

2. Craig Sinclair's descriptions of her mind-reading technique in *Mental Radio* echo those of Gurney's percipients, but they are far richer in insight. The most essential quality for telepathy, she claims, is "undivided attention or concentration," and she depicts a highly active, controlling process. "It isn't thinking," she explains, "it is inhibiting thought." She tells readers to concentrate on a simple thought, suppressing all "memory trains" associated with it, all thoughts about it, and all environmental or sensory inputs. One must achieve a state that is like sleep, but without ever falling asleep, a "passive state of mind and body." At last, she explains, one "gives the order to the unconscious mind to tell you what is on the paper" (Sinclair 178–87).

3. In "Wireless" (1902), Rudyard Kipling playfully juxtaposed transmissions received on a wireless telegraph with transmissions received by a shop clerk—directly from the mind of deceased poet John Keats. See Jill Galvan, "Female Channeling: Women, Technology, and the Occult 1880–1915."

4. Craig Sinclair, Upton Sinclair's clairvoyant wife, believed that we have "several mental entities," some of which were usually asleep at any given time. In teaching readers how to receive transmitted impressions, she advised them to give orders to their subconscious minds "as if talking directly to another self" (Sinclair 180, 187).

5. The editor quoted uses Twain's real name, Samuel Clemens.

6. In 1911, Barrett would hope for "a far more perfect interchange of thought than by the clumsy mechanism of speech" (Barrett 69). The phrasing of both writers reflects the language of reports to the Society for Psychical Research throughout the 1880s and 1890s.

7. I agree with Regenia Gagnier's assessment that because of Stoker's theater job, "management and information technique and technology" were "what he knew best" (152).

8. Many critics have commented on the patchwork structure of *Dracula*. Franco Moretti calls the novel "a network of letters, diaries, notes, [and] telegrams . . . [in which] the description and ordering of events is reserved for the British alone" (437). Carol Senf believes that by incorporating letters and newspaper clippings, "the very written record of everyday life," Stoker deemphasizes the story's mythological aspects. The use of multiple perspectives and official-sounding correspondence does not make this written record any more believable, however, since many of the narrators are unreliable (Senf 422).

9. Rebecca Pope writes that "*Dracula* stresses the process of textual production as much as it does textual product" (202).

10. I agree with Kathleen L. Spencer's suggestion that both literary and scientific narratives rely on the same chronological ordering techniques: "Only with chronology does narrative emerge; only then does a collection of data turn into a hypothesis" (220).

11. In an analysis of Mina's "textual work as textile work," Rebecca Pope compares Stoker's character to female weavers and narrators in several cultures' mythologies. Like Sheherazade, Mina is telling a story to save her life. She also has some ties to Philomela, who was raped by King Tereus and then wove the image into a tapestry when he cut out her tongue so that she could not talk. Mina bears some resemblance to Arachne as well, who wove a tapestry of women "appropriated" by male gods and was duly punished for it. It would be "critically reductive," argues Pope, to see these weaving comparisons as "trivializing" to Mina, for in all of these legends, women's weaving and narration are empowering (212–13).

12. Several scholars have pointed out the parallels between Dracula's and the hunters' writing. Rebecca Pope's study, "Writing and Biting in *Dracula*," explores the pivotal role of language in the novel, and Friedrich Kittler's *Discourse Networks* compares the impressions made by Mina's typewriter keys to the indentations Dracula leaves on Mina's throat (355–56). Stoker was probably conscious of this parallel, for his phrasing suggests an ongoing preoccupation with the idea. The first time he ever saw Henry Irving, Stoker described him as "a figure . . . whose ridicule seemed to *bite*" (Farson 24, original emphasis). In *Dracula*, Mina's comment that "you can't go on for some years teaching etiquette and decorum to other girls without the pedantry of it biting into yourself a bit" suggests an association of biting with the oppressive forces of male power (155). Like her stays, these forces restrict her movements and bite into her body.

13. Rosemary Jann writes that in *Dracula*, "reportage is clearly intended as an attempt at rational control over experience" (279). Rebecca Pope points out that

Stoker is indebted to Richardson for the idea of "writing 'to the moment' and under threat" (208).

14. While Stoker's characters rejoice in the way that language informs and consoles, Stoker's novel also offers innumerable examples of language's weaknesses. Like Eliot and James, Stoker depicts written and spoken language as a noisy, self-propagating, and highly flawed means of communication. When Jonathan Harker receives a note saying, "Sam Bloxam, Korkrans, 4, Poters Cort, Bartel Street, Walworth. Arsk for the depite," he experiences some difficulty following its author's directions (231). *Dracula* is a polyglot novel, and while its dialects and malapropisms were probably intended as comic relief—at the expense of the lower classes—, they also reveal a great deal about the way the main characters communicate (Pope 200). Like Harker, who can find no "depite," the vampire hunters read literally, responding to signs in a thoughtless, machinelike way. Van Helsing's mangled English is a running joke throughout the novel, and despite his purported knowledge and language skills, one can see that he understands as poorly as he speaks. Trying to learn of Dracula's escape by sea, the "open-minded" professor can make nothing of an angry sea captain's talk of "bloom and blood" (275). Like Sancho Panza, Van Helsing unknowingly alters common expressions, illustrating language's capacity to create new meanings when it is randomly rearranged. His statement that "the milk that is spilt cries not out afterwards" resonates in an interesting way with Stoker's description of Mina's oral rape: "a terrible resemblance to a child forcing a kitten's nose into a saucer of milk to compel it to drink" (208, 247). A random generator, Van Helsing's mind shows the possibilities of the medium he has yet to master (Kittler 206–12).

15. Garrett Stewart calls Mina's typing "a parody of reproduction . . . a travesty . . . of the Count's own self-multiplying practices" (9).

16. For an analysis of how the keyboard created a new "discourse network," see Friedrich Kittler's 1985 study. Kittler argues that the typewriter altered our concept of language from one of a natural, poetic medium learned from the mother's mouth to one of a randomly generated, self-propagating, machinelike babble. Stoker's novel supports Kittler's claim that the typewriter changes the feel of writing. After working with a keyboard, Mina reflects, "I should have felt quite astray doing the work if I had to write with a pen" (303).

17. According to Stephen Arata, Harker's "obsession with trains running on time" and description of crossing a boundary are typical of Victorian travel narratives about Eastern Europe. Such accounts could make sense of the East only by presuming that there was an "unbridgeable gap" between East and West (635–36).

18. Rosemary Jann describes timetables as a "graphic device for ordering experience" (279).

19. In Regenia Gagnier's description, "England casts the net that catches the Count and defeats the forces of myth and superstition by enlisting an international network of scientists and scholars . . . [the] international networks of law, business, and government" (149).

20. According to Rebecca Pope, the hunters follow a paradigm that stresses "the

writing and reading of texts as a way of producing knowledge" (199). Regenia Gagnier concurs and argues that Dracula's way of knowing (based on evolution and corporeal memory) is the antithesis of the hunters', which is based on recorded information. She views these systems of intelligence as "two different descriptions of order within disorder" and believes that they compete throughout the novel (146). I agree with the distinction but would argue that these two ways of knowing are closely related in the novel, often close to analogous. If they are opposites, they are mirror images of each other.

21. Critics have read the male bonding in *Dracula* in a number of ways. Rebecca Pope points out the characters' references to telling stories around a campfire, seeing "a communal bonding through the exchange of narratives. " She also believes that their transfusions to Lucy show "males bonding with each other in and through a woman" (207). Garret Stewart notes the resemblance to idealized scenes of Victorian families reading around their hearths and calls the hunters "an extended family of data gatherers" (8–9).

22. Kathleen Spencer points out a further reason for Victorian approval of networking: "for the Victorians, solitude greatly increased sexual danger" (215).

23. In a final note, Stoker's fictitious editor states that "in all the mass of material of which the record is composed, there is hardly one authentic document" (326). Regenia Gagnier believes that this lack of authenticity refers to the absence of Dracula's own story, the missing original of which the hunters' record is a mere copy (147).

24. In her reading of *Dracula,* Carol Senf argues that the novel is about the similarities between evil and good. "The only difference between Dracula and his opponents," she writes, "is the narrators' ability to state individual desire in terms of what they believe is a common good" (427).

25. Critics disagree considerably about the extent to which Dracula has mastered Western ways. Stephen Arata calls Dracula an "Occidentalist," arguing that his study and colonization of England mirror British Orientalism and imperialism. "Before Dracula successfully invades the spaces of his victims' bodies or land," he writes, "he first invades the spaces of their knowledge" (634). Regenia Gagnier, however, finds Dracula's knowledge of British customs "amateurish" (151). Rosemary Jann echoes Van Helsing's low assessment of Dracula's mind, writing that the Count lacks the "tools of deductive logic" and "can only leap from point to point, without power of generalization" (281–82). I agree with Arata that Dracula mimics and outperforms the British through much of the novel.

26. Regenia Gagnier believes that the Count "expresses what the men in the novel . . . want" but sees the hunters' information technology and "discourse of machinery" as "antagonistic to everything the Count represents" (142). Both the Count and his pursuers network in strikingly similar ways, however.

27. When one considers Dracula's relationship to time, one sees the wisdom of Regenia Gagnier's argument that he is a creature of evolution, not information (146). Stephen Arata agrees that because Dracula is immortal, his growth and development can be seen as that of a species or race (640).

28. Van Helsing is probably referring to electrophysiological experiments conducted by Gustav Fritsch, Edouard Hitzig, and David Ferrier. German physiologists Fritsch and Hitzig, who studied with DuBois-Reymond, demonstrated in 1870 that one could produce movements in an anesthetized animal by electrically stimulating its brain. So consistent were their results that in the motor cortex, they could map out the territories corresponding to body parts (Brazier, *Neurophysiology in the Nineteenth Century* 155–58, 165–70). British physiologist David Ferrier, who also used electricity to map the brain, was tried for vivisection in the early 1880s and was strongly criticized in the press. It is most likely such coverage of electrophysiological studies that leads Van Helsing to call them "unholy."

29. Regenia Gagnier describes Dracula as having a "bodily intelligence," and Jennifer Wicke writes that the telegraph is "equivalent to the telepathic, telekinetic communication Dracula is able to have with Mina" (Gagnier 155; Wicke, "Vampiric Typewriting" 475).

30. Rosemary Jann proposes that Dracula is an "evil double" of Van Helsing because of their common ability to hypnotize (276). In a more detailed comparison, Garrett Stewart writes that Dracula's telepathy makes him the "necromantic counterpart of the new techniques of telegraphy and phonology." Stewart argues that Van Helsing's hypnotism, a kind of reading, becomes the passive analog of Dracula's active hypnotism, which can project thoughts (10).

31. Rebecca Pope compares Mina to a letter that "carries information between two sets of male minds" (203). In Garrett Stewart's description, she is a "living dictaphone," a "telepathic chiasm," and a "hypnotic switch-point" (12–13). Jennifer Wicke's argument that "communication flows through" late nineteenth-century women, "telegraphically or otherwise enhanced," is as true for *Dracula* as for Henry James's "In the Cage," which appeared a year later (Wicke, "Henry James's Second Wave" 148).

32. The original French reads: "Celui qui n'a fonction que de manger commande" (39). My translation.

33. Rebecca Pope offers another reason for Dracula's parasitism, claiming that the novel is parasitic by nature because it "lives off of other discourses" (199).

34. Franco Moretti proposes that Dracula represents capital itself, growing endlessly at the expense of humanity (432).

35. Judith Halberstam observes that in the novel, "vampirism somehow interferes with the natural ebb and flow of currency" and proposes that Dracula's dusty heaps of gold violate the capitalist demand that "money . . . should be used and circulated" (346).

Bibliography

Andres, Sophia. "The Germ and the Picture in *Middlemarch.*" *English Literary History* 55 (1988): 853–68.

Arata, Stephen D. "The Occidental Tourist: *Dracula* and the Anxiety of Reverse Colonization." *Victorian Studies* 33 (1990): 621–45.

Asmann, Edwin N. *The Telegraph and the Telephone: Their Development and Role in the Economic History of the United States: The First Century 1844–1944.* Lake Forest, IL: Lake Forest College, 1980.

Aswell, E. Duncan. "James's *In the Cage:* The Telegraphist as Artist." *Texas Studies in Literature and Language* 8 (1966): 375–84.

Babbage, Charles. *The Works of Charles Babbage.* Edited by Martin Campbell-Kelly. 10 vols. New York: New York University Press, 1989.

Balsamo, Anne. "Feminism for the Incurably Informed." In *Flame Wars: The Discourse of Cyberculture,* edited by Mark Dery. Durham: Duke University Press, 1994.

Barnard, Charles. "Kate: An Electro-Mechanical Romance." In *Lightning Flashes and Electric Dashes: A Volume of Choice Telegraphic Literature, Humor, Fun, Wit, and Wisdom,* edited by William John Johnston. New York: Johnston, 1877.

Barrett, W. F. *Psychical Research.* New York: Holt; London: Williams and Norgate, 1911.

Barrett, W. F., Edmund Gurney, and F. W. H. Myers. "First Report of the Committee on Thought Reading." *Proceedings of the Society for Psychical Research* 1 (1882–83): 13–34.

Bauer, Dale M., and Andrew Lakritz. "Language, Class, and Sexuality in Henry James's 'In the Cage.'" *New Orleans Review* 14.3 (1987): 61–69.

Beer, Gillian. "Circulatory Systems: Money and Gossip in *Middlemarch.*" *Cahiers Victoriens et Edouardiens* 26 (October 1987): 47–62.

———. *Darwin's Plots: Evolutionary Narrative in Darwin, George Eliot, and Nineteenth-Century Fiction.* Second Edition. Cambridge: Cambridge University Press, 2000.

Bernard, Claude. *Leçons sur la physiologie et la pathologie du système nerveux.* 2 vols. Paris: Baillière, 1858.

Bichat, Xavier. *General Anatomy.* Translated by Constance Coffyn. London: N.p., 1824.

Brazier, Mary A. B. *A History of Neurophysiology in the Nineteenth Century.* New York: Raven Press, 1988.

————. *A History of Neurophysiology in the Seventeenth and Eighteenth Centuries.* New York: Raven Press, 1984.

Brehm, A. E. *Illustriertes Thierleben: Eine allgemeine Kunde des Thierreichs.* 6 vols. Bilburghausen: Verlag des Bibliographischen Instituts, 1869.

Briggs, Charles F., and Augustus Maverick. *The Story of the Telegraph and a History of the Great Atlantic Cable.* New York: Rudd and Carleton, 1858.

Brody, Selma. "Origins of George Eliot's 'Pier-Glass' Image." *English Language Notes* 22.2 (1984): 55–58.

————. "Physics in *Middlemarch:* Gas Molecules and Ethereal Atoms." *Modern Philology* 85.1 (1987): 42–53.

Buxton, H. W. *Memoir of the Life and Labors of the Late Charles Babbage, Esq.* The Charles Babbage Institute Reprint Series for the History of Computing. Cambridge: MIT Press, 1988.

Campbell, Lewis, and William Garnett. *The Life of James Clerk Maxwell.* London: Macmillan, 1882.

Carey, James W. *Communication as Culture: Essays on Media and Society.* Boston: Unwin Hyman, 1989.

Chladni, Ernst Florens Friedrich von. *Die Akustik.* Leipzig: Breitkopf and Hartel, 1802.

————. *Theorie des Klanges.* Leipzig: Weidmanns, 1787.

Churchill, L. A. "Playing with Fire." In *Lightning Flashes and Electric Dashes: A Volume of Choice Telegraphic Literature, Humor, Fun, Wit, and Wisdom,* edited by William John Johnston. New York: Johnston, 1877.

Clark, Latimer. "Memoir of Sir William Fothergill Cooke." In *Extracts from the Private Letters of the Late Sir William Fothergill Cooke, 1836–39, Relating to the Invention and Development of the Electric Telegraph,* by William Fothergill Cooke. Edited by F. H. Webb. London: E. and F. N. Spon, 1895.

Clarke, Edwin, and L. S. Jacyna. *Nineteenth-Century Origins of Neuroscientific Concepts.* Berkeley: University of California Press, 1987.

Clarke, Edwin, and C. D. O'Malley. *The Human Brain and Spinal Cord: An Historical Study Illustrated by Writings from Antiquity to the Twentieth Century.* 2d ed. San Francisco: Norman, 1996.

Cooke, William Fothergill. *Extracts from the Private Letters of the Late Sir William Fothergill Cooke, 1836–39, Relating to the Invention and Development of the Electric Telegraph.* Edited by F. H. Webb. London: E. and F. N. Spon, 1895.

Cosslett, Tess. *The "Scientific Movement" and Victorian Literature.* New York: St. Martin's Press, 1982.

Cross, J. W. *George Eliot's Life as Related in Her Letters and Journals.* 3 vols. New York: Harper, 1903.

Davidson, Cathy, and Linda Wagner-Martin, eds. *The Oxford Companion to Women's Writing in the United States.* Oxford: Oxford University Press, 1995.

Dery, Mark. *Flame Wars: The Discourse of Cyberculture.* Durham: Duke University Press, 1994.

Diderot, Denis. *Eléments de Physiologie.* Edited by Jean Mayer. Paris: Marcel Didier, 1964.

———. *Le Rêve de D'Alembert*. Edited by Paul Vernière. Paris: Marcel Didier, 1951.

Dierig, Sven. "Hirngespinste am Klavier: Über 'chopinisierte' Nervensaiten im Berliner Fin de Siècle." In *Ecce Cortex: Beiträge zur Geschichte des modernen Gehirns*, edited by Michael Hagner. Göttingen: Wallstein, 1999.

Dosch, Hans Günther. "The Concept of Sign and Symbol in the Work of Hermann Helmholtz and Heinrich Hertz." *Etudes de lettres* 1–2 (1997): 47–61.

Doyle, Arthur Conan. *The History of Spiritualism*. 2 vols. New York: George H. Doran, 1926. Reprint, New York: Arno, 1975.

DuBois-Reymond, Emil. *On Animal Electricity: Being an Abstract of the Discoveries of Emil DuBois-Reymond*. Edited and translated by Henry Bence Jones. London: Churchill, 1852.

———. "On the Time Required for the Transmission of Volition and Sensation through the Nerves." In *Croonian Lectures on Matter and Force*, edited by Henry Bence Jones. London: Churchill, 1868.

———. *Reden*. 2 vols. Leipzig: Veit, 1887.

Edel, Leon. *Henry James: A Life*. New York: Harper and Row, 1985.

The Electric Telegraph Company Chart of the Company's Telegraphic System in Great Britain, 1853.

"Electricity and the Electrical Telegraph." *The Cornhill Magazine* 2 (1860): 61–73.

Eliot, George. *The George Eliot Letters*. Edited by Gordon S. Haight. 9 vols. New Haven: Yale University Press, 1955.

———. "The Influence of Rationalism." *Fortnightly Review* 1 (1865): 46.

———. *Middlemarch*. 1871–72. New York: Norton, 1977.

———. "The Natural History of German Life." In *Middlemarch*. By George Eliot. Edited by Bert G. Hornback. New York: Norton, 1977.

Eros, Paul. "Interassured of the Mind STOP: Telegraphy and the Question of Identity in Edith Wharton and Henry James." Paper presented at the Northeast Modern Language Association Conference, Pittsburgh, Penn., 17 April, 1999.

Faraday, Michael. *Experimental Researches in Electricity*. 2 vols. 1839–55. Reprint, New York: Dover Publications, 1965.

Farson, Daniel. *The Man Who Wrote* Dracula: *A Biography of Bram Stoker*. New York: St. Martin's Press, 1975.

Feimer, Joel N. "Bram Stoker's *Dracula:* The Challenge of the Occult to Science, Reason, and Psychiatry." In *Contours of the Fantastic: Selected Essays from the Eighth International Conference on the Fantastic in the Arts*, edited by Michelle K. Langford. Westport, Conn.: Greenwood, 1987.

Fellows, Otis. *Diderot*. Boston: Twayne, 1977.

Feltes, N. N. "George Eliot's 'Pier-Glass': The Development of a Metaphor." *Modern Philology* 67 (August 1969): 69–71.

Forrester, John. "Lydgate's Research Project in *Middlemarch*." *George Eliot–George Henry Lewes Newsletter* 16–17 (September 1990): 2–6.

Foucault, Michel. *Discipline and Punish: The Birth of the Prison*. Translated by Alan Sheridan. New York: Vintage-Random House, 1979.

Francis, John. *A History of the English Railway: Its Social Relations 1820–1845.* 2 vols. London: Longman, 1851.

"Freaks of the Telegraph." *Blackwood's Magazine* 129 (1881): 468–78.

Fullinwider, S. P. "Hermann von Helmholtz: The Problem of Kantian Influence." *Studies in History and Philosophy of Science* 21.1 (1990): 41–55.

Gabler-Hover, Janet. "The Ethics of Determinism in Henry James's 'In the Cage.'" *Henry James Review* 13 (1992): 253–74.

Gagnier, Regenia. "Evolution and Information, or Eroticism and Everyday Life, in *Dracula* and Late Victorian Aestheticism." In *Sex and Death in Victorian Literature,* edited by Regina Barreca. Bloomington: Indiana University Press, 1990.

Galvan, Jill. "Feminine Channeling: Women, Technology, and the Occult 1880–1915." Ph.D. diss., UCLA, in progress.

Galvani, Luigi. *Commentary on the Effect of Electricity on Muscular Motion.* Translated by Robert Montraville Green. 1791. Reprint, Cambridge, Mass.: Elizabeth Licht, 1953.

———. *De viribus electricitatis in motu musculari.* Bononia: Instituti Scientarium, 1791.

Gerber, John C. *Mark Twain.* Boston: Twayne, 1988.

Gerlach, Josef von. "Ueber die Structur der grauen Substanz des menschlichen Grosshirns." *Centralblatt für die medicinischen Wissenschaften* 18 (1872): 273–75.

Gillispie, Charles Coulston. *The Edge of Objectivity: An Essay in the History of Scientific Ideas.* Princeton: Princeton University Press, 1960.

———, ed. *Dictionary of Scientific Biography.* 15 vols. New York: Scribner, 1970.

Glover, David. "Bram Stoker and the Crisis of the Liberal Subject." *New Literary History* 23 (1992): 983–1002.

Golgi, Camillo. "La doctrine du neurone." In *Les Prix Nobel en 1906.* Edited by M. C. G. Santesson. Stockholm: Imprimerie Royale, 1908.

———. "Recherches sur l'histologie des centres nerveux." *Archives italiennes de biologie* 3 (1883): 285–317; 4 (1884): 92–123.

———. "La rete nervosa diffusa degli organi centrali del sistema nervoso su significato fisiologico." 1891. *Opera omnia.* 4 vols. Milan: Ulrico Hoepli, 1903.

———. "Sulla fina anatomia degli organi centrali del sistema nervoso." 1883. In *Opera omnia.* 4 vols. Milan: Ulrico Hoepli, 1903.

Greenberg, Robert A. "Plexuses and Ganglia: Scientific Allusion in *Middlemarch.*" *Nineteenth-Century Fiction* 30 (1975–76): 33–52.

Greenway, John L. "Seward's Folly: *Dracula* as a Critique of 'Normal Science.'" *Stanford Literature Review* 3 (1986): 213–30.

Gurney, Edmund, Frederic W. H. Myers, and W. F. Barrett. "Second Report of the Committee on Thought-Transference." *Proceedings of the Society for Psychical Research* 1 (1882–83): 70–98.

Gurney, Edmund, Frederic W. H. Myers, and Frank Podmore. *Phantasms of the Living.* 2 vols. London: Trübner, 1886.

Hagner, Michael. "Les Frères ennemis de la physiologie." *Les Cahiers de science et vie* 58 (2000): 38–45.

———. *Homo cerebralis: Der Wandel vom Seelenorgan zum Gehirn.* Frankfurt: Insel, 1997.

Haight, Gordon S. *George Eliot: A Biography.* London: Penguin, 1992.

Halberstam, Judith. "Technologies of Monstrosity: Bram Stoker's *Dracula.*" *Victorian Studies* 36 (1993): 333–52.

Haller, Albertus. *First Lines of Physiology.* Edited and translated by William Cullen. 1786. Reprint, New York: Johnson, 1966.

Haraway, Donna. *Modest_Witness@SecondMillenium.FemaleManc_MeetsOncomouseTM : Feminism and Technoscience.* New York: Routledge, 1997.

Hayles, N. Katherine. *Chaos Bound: Orderly Disorder in Contemporary Literature and Science.* Ithaca: Cornell University Press, 1990.

———. "Escape and Constraint: Three Fictions Dream of Moving from Energy to Information." Working paper. 1 February, 1999.

———. *How We Became Posthuman: Virtual Bodies in Cybernetics, Literature, and Informatics.* Chicago: University of Chicago Press, 1999.

Heimann, P. M. "Helmholtz and Kant: The Metaphysical Foundations of *Über die Erhaltung der Kraft.*" *Studies in History and Philosophy of Science* 5 (1974): 205–38.

Helmholtz, Hermann von. "Note sur la vitesse de propagation de l'agent nerveux dans les nerfs rachidiens." *Comptes rendus de L'Académie des Sciences* 30 (1850): 204–6; 33 (1851): 262–65.

———. *Science and Culture: Popular and Philosophical Essays.* Edited by David Cahan. Chicago: University of Chicago Press, 1995.

———. "Vorläufiger Bericht über die Fortpflanzungsgeschwindigkeit der Nervenreizung." *Archiv für Anatomie, Physiologie, und wissenschaftliche Medizin* 17 (1850): 71–73, 276–364.

———. *Vorlesungen über theoretische Physik.* Leipzig: Barth, 1903.

D'Holbach, Paul. *Système de la nature.* 2 vols. Hildesheim: Olms, 1966.

Hooton, Joy W. "*Middlemarch* and Time." *English Studies* 52 (1971): 188–202.

Hutchinson, Stuart. "James's *In the Cage*: A New Interpretation." *Studies in Short Fiction* 19.1 (1982): 19–25.

Hyman, Anthony. Introduction to *Memoir of the Life and Labors of the Late Charles Babbage, Esq.*, by H. W. Buxton. The Charles Babbage Institute Reprint Series for the History of Computing. Cambridge: MIT Press, 1988.

James, Henry. *Eight Tales from the Major Phase: "In the Cage" and Others.* Edited by Morton Dauwen Zabel. New York: Norton, 1958.

James, Henry, and William James. *Selected Letters.* Edited by Ignas K. Skrupskelis and Elizabeth M. Berkeley. Charlottesville: University Press of Virginia, 1997.

Jann, Rosemary. "Saved by Science? The Mixed Messages of Stoker's *Dracula.*" *Texas Studies in Literature and Language* 31 (1989): 273–87.

Johnston, William John, ed. *Lightning Flashes and Electric Dashes: A Volume of Choice Telegraphic Literature, Humor, Fun, Wit, and Wisdom.* New York: Johnston, 1877.

Jumeau, Alain. "Les Premiers chemins de fer, la réalité et la fiction: Fanny Kemble

(*Record of a Girlhood*), Samuel Smiles (*Lives of the Engineers*) et George Eliot (*Middlemarch*)." *Etudes anglaises* 43 (1990): 403–13.

Ketterer, David. Introduction to *The Science Fiction of Mark Twain,* edited by David Ketterer. Hamden, Conn.: Archon-Shoestring Press, 1984.

Kieve, Jeffrey. *The Electric Telegraph: A Social and Economic History.* Newton Abbot: David and Charles, 1973.

Kipling, Rudyard. "Wireless." In *Traffics and Discoveries.* [1904]. N.p.: Doubleday, Page and Company, 1914.

Kitchell, Anna Teresa. Quarry *for* Middlemarch. Berkeley: University of California Press, 1950.

Kittler, Friedrich A. *Discourse Networks, 1800/1900.* Translated by Michael Metteer. Foreword by David E. Wellbery. Stanford: Stanford University Press, 1990.

Lakoff, George, and Mark Johnson. *Metaphors We Live By.* Chicago: University of Chicago Press, 1980.

Lawrence, Christopher. "The Nervous System and Society in the Scottish Enlightenment." In *Natural Order: Historical Studies of Scientific Culture,* edited by Barry Barnes and Steven Shapin. London: Sage, 1979.

Lenoir, Timothy. "The Eye as Mathematician: Clinical Practice, Instrumentation, and Helmholtz's Construction of an Empiricist Theory of Vision." In *Hermann von Helmholtz and the Foundations of Nineteenth-Century Science,* edited by David Cahan. Berkeley: University of California Press, 1993.

———. "Helmholtz and the Materialities of Communication." *Osiris* 9 (1994): 185–207.

———. *Inscribing Science: Scientific Texts and the Materiality of Communication.* Stanford: Stanford University Press, 1998.

———. "Models and Instruments in the Development of Electrophysiology, 1845–1912." *Historical Studies in the Physical and Biological Sciences* 17.1 (1986): 1–54.

Lewes, George Henry. *The Life of Goethe.* London: Smith and Elder, 1855; New York: Frederick Ungar, 1965.

———. *The Physical Basis of Mind.* London: Trübner, 1877.

Lorenz, Paul H. "Technology and Development: Opposition to the Railway in *Middlemarch.*" *George Eliot Fellowship Review* 22 (1991): 21–23.

Maclachlan, J. M. "A Perilous Christmas Courtship; or, Dangerous Telegraphy." In *Lightning Flashes and Electric Dashes: A Volume of Choice Telegraphic Literature, Humor, Fun, Wit, and Wisdom,* edited by William John Johnston. New York: Johnston, 1877.

Marvin, Carolyn. *When Old Technologies Were New: Thinking about Electric Communication in the Late Nineteenth Century.* New York: Oxford University Press, 1988.

Maxwell, James Clerk. "Reflex Musings: Reflections from Various Surfaces." In *The Life of James Clerk Maxwell.* By Lewis Campell and William Garnett. London: MacMillan, 1882.

McCallum, Ellen. "Love Lines, Crossed Wires: Telegraphy and the Instability of Identity." Working paper. 14 September, 1998.

McCarthy, Patrick J. "Lydgate, 'The New, Young Surgeon' of *Middlemarch.*" *Studies in English Literature* 10 (1970): 805–16.

McCormack, Kathleen. "George Eliot and the Pharmakon: Dangerous Drugs for the Condition of England." *Victorians Institute Journal* 14 (1986): 35–51.

McMaster, Juliet. "'A Microscope Directed on a Water-Drop': Chapter Nineteen." In *Approaches to Teaching George Eliot's* Middlemarch. Edited by Kathleen Blake. New York: Modern Language Association, 1990.

Menke, Richard. "Fiction as Vivisection: G. H. Lewes and George Eliot." *English Literary History* 67 (2000): 617–53.

———. "Telegraphic Realism: Henry James's *In the Cage.*" *PMLA* 115 (2000): 975–90.

Mertens, Joost. "Shocks and Sparks: The Voltaic Pile as a Demonstration Device." *Isis* 89 (1998): 300–311.

la Mettrie, Julien Offray de. *La Mettrie's L'Homme Machine: A Study in the Origins of an Idea.* Edited by Aram Vartanian. Princeton: Princeton University Press, 1960.

Miller, J. Hillis. "Optic and Semiotic in *Middlemarch.*" In *The Worlds of Victorian Fiction,* edited by Jerome H. Buckley. Cambridge: Harvard University Press, 1975.

———. "The Roar on the Other Side of Silence: Otherness in *Middlemarch.*" *Edda* 3 (1995): 237–45.

Mitchell, S. Weir. *Autobiography.* Papers, College of Physicians, Philadelphia, Penn.

Miss X. "Apparent Sources of Subliminal Messages." *Proceedings of the Society for Psychical Research* 11 (1895): 114–17.

Moody, Andrew J. "'The Harmless Pleasure of Knowing': Privacy in the Telegraph Office and Henry James's 'In the Cage.'" *Henry James Review* 16.1 (1995): 53–65.

Moretti, Franco. "A Capital *Dracula.*" In *Dracula: A Norton Critical Edition,* edited by Nina Auerbach and David J. Skal. New York: Norton, 1997.

Morse, Samuel F. B. *Samuel F. B. Morse: His Letters and Journals.* Edited by Edward Lind Morse. 2 vols. Boston: Houghton Mifflin, 1914.

Moseley, Maboth. *Irascible Genius: A Life of Charles Babbage, Inventor.* London: Hutchinson, 1964.

Mullaly, John. *The Laying of the Cable, or The Ocean Telegraph.* New York: Appleton, 1858.

Müller, Johannes. *Elements of Physiology.* Translated by William Baly. 2d ed. 2 vols. London: Taylor and Walton, 1839.

Myers, F. W. H. "On Telepathic Hypnotism and Its Relation to Other Forms of Hypnotic Suggestion." *Proceedings of the Society for Psychical Research* 4 (1886–87): 127–28.

———. "The Subliminal Consciousness." *Proceedings of the Society for Psychical Research* 7 (1890–91).

The New Encyclopedia Britannica. 15th ed. 29 vols. Chicago: University of Chicago Press, 1987.

Nietzsche, Friedrich. "Ueber Wahrheit und Lüge im aussermoralischen Sinne." *Kri-*

tische Studienausgabe. Edited by Giorgio Colli and Mannino Montinari. 15 vols. München: de Gruyter, 1988.

Norrman, Ralf. "The Intercepted Telegram Plot in Henry James's 'In the Cage.'" *Notes and Queries* 24 (1977): 425–27.

O'Brien, Patrick. *Railways and the Economic Development of Western Europe, 1830–1914.* New York: St. Martin's Press, 1983.

Olesko, Kathryn M., and Frederic L. Holmes. "Experiment, Quantification, and Discovery: Helmholtz's Early Physiological Researches, 1843–50." In *Hermann von Helmholtz and the Foundations of Nineteenth-Century Science,* edited by David Cahan. Berkeley: University of California Press, 1993.

di Pasquale, P., Jr. "The Imagery and Structure of *Middlemarch.*" *Southern Review* 13 (1980): 425–35.

Pater, Walter. "The Renaissance." In *Middlemarch.* By George Eliot. Edited by Bert G. Hornback. New York: Norton, 1977.

Pera, Marcello. *The Ambiguous Frog: The Galvani-Volta Controversy on Animal Electricity.* Translated by Jonathan Mandelbaum. Princeton: Princeton University Press, 1992.

Perkowitz, Sidney. "Connecting with E. M. Forster." *The American Prospect* 7.26 (1996): 86–89.

Pope, Rebecca A. "Writing and Biting in *Dracula.*" *LIT* 1 (1990): 199–216.

Pratt, John Clark, and Victor A. Neufeldt. *George Eliot's* Middlemarch *Notebooks: A Transcription.* Berkeley: University of California Press, 1979.

Prescott, George B. *History, Theory, and Practice of the Electric Telegraph.* Boston: Ticknor and Fields, 1860.

Pupilli, Giulio C. Introduction to *Commentary on the Effect of Electricity on Muscular Motion,* by Luigi Galvani. Edited by Robert Montraville Green. Cambridge, Mass.: Elizabeth Licht, 1953.

Ramón y Cajal, Santiago. "Algunas conjeturas sobre el mecanismo anatómico de la ideación, asociación, y atención." *Revista de medicina y cirugía prácticas* (1895): 3–14.

———. "Conexión general de los elementos nerviosos." 1889. In *Trabajos escogidos 1880–90.* Madrid: N.p., 1924.

———. "The Croonian Lecture: La Fine Structure des Centres Nerveux." *Proceedings of the Royal Society* 55 (1894): 444–68.

———. *Elementos de histología normal y de técnica micrográfica.* Madrid: Moya, 1904.

———. "Estructura de los centros nerviosos de las aves." 1888. In *Trabajos escogidos 1880–90.* Madrid: N. p., 1924.

———. "¿Neuronismo o reticularismo? Las pruebas objetivas de la unidad anatómica de las células nerviosas." *Archivos de neurobiología* 13 (1933): 219–91.

———. "Structure et Connexions des Neurones." In *Les Prix Nobel 1906.* Edited by M. C. G. Santesson. Stockholm: Imprimerie Royale, 1908.

———. *Textura del sistema nervioso del hombre y de los vertebrados.* Madrid: Moya, 1899.

Ranvier, Louis. *Technisches Lehrbuch der Histologie*. Translated by W. Nicati and H. von Wyss. Leipzig: Vogel, 1877.

Romanes, George John. *Jelly-Fish, Star-Fish, and Sea Urchins: Being a Research on Primitive Nervous Systems*. London: Kegan Paul, 1885.

Ronalds, Francis. *Descriptions of an Electrical Telegraph and of Some Other Electrical Apparatus*. London: Hunter, 1823.

Rothfield, Lawrence. *Vital Signs: Medical Realism in Nineteenth-Century Fiction*. Princeton: Princeton University Press, 1992.

Sabine, Robert. *The Electric Telegraph*. London: Virtue, 1867.

Salzberg, Joel. "Mr. Mudge as Redemptive Fate: Juxtaposition in James's *In the Cage*." *Studies in the Novel* 11 (1979): 63–76.

Sauer, George. *The Telegraph in Europe*. Paris: N.p., 1869.

Schivelbusch, Wolfgang. *The Railway Journey: The Industrialization of Time and Space in the Nineteenth Century*. Berkeley: University of California Press, 1986.

Schlechta, Karl, and Anni Anders. *Friedrich Nietzsche: Von den verborgenen Anfängen seines Philosophierens*. Stuttgart: Friedrich Frommann, 1962.

Seltzer, Mark. *Bodies and Machines*. New York: Routledge, 1992.

Senf, Carol A. "*Dracula*: The Unseen Face in the Mirror." In *Dracula: A Norton Critical Edition*, edited by Nina Auerbach and David J. Skal. New York: Norton, 1997.

Serres, Michel. *Le Parasite*. Paris: Bernard Grasset, 1980.

Shuttleworth, Sally. *George Eliot and Nineteenth-Century Science: The Make Believe of a Beginning*. Cambridge: Cambridge University Press, 1984.

Siemens, Werner von. *Werner von Siemens, Inventor and Entrepreneur: Recollections of Werner von Siemens*. London: Lund, 1966.

Simmons, Jack. *The Victorian Railway*. New York: Thames and Hudson, 1991.

Sinclair, Upton. *Mental Radio*. Pasadena: Station A, 1930.

Smiles, Samuel. *The Life of George Stephenson, Railway Engineer*. 5th ed. London: Murray, 1858.

Spencer, Kathleen L. "Purity and Danger: *Dracula*, the Urban Gothic, and the Late Victorian Degeneracy Crisis." *English Literary History* 59 (1992): 197–225.

Standage, Tom. *The Victorian Internet: The Remarkable Story of the Telegraph and the Nineteenth Century's Online Pioneers*. New York: Walker, 1998.

Stewart, Garrett. "'Count Me In': *Dracula*, Hypnotic Participation, and the Late-Victorian Gothic of Reading." *LIT* 5.1 (1994): 1–18.

Stoker, Bram. *Dracula: A Norton Critical Edition*. Edited by Nina Auerbach and David J. Skal. New York: Norton, 1997.

———. "Notes for Dracula." In *Dracula: A Norton Critical Edition*. Edited by Nina Auerbach and David J. Skal. New York: Norton, 1997.

Stone, Allucquère Rosanne. *The War of Desire and Technology at the Close of the Mechanical Age*. Cambridge: MIT Press, 1995.

Stone, William B. "On the Background of James's *In the Cage*." *American Literary Realism* 6 (1973): 243–47.

Stubbs, Katherine. "Mechanizing the Female: Discourse and Control in the Industrial Economy." In *The Image of Technology in Literature, the Media, and Soci-*

ety, edited by Will Wright and Steve Kaplan. Pueblo: University of Southern Colorado Press, 1994.

Stwertka, Eve Marie. "The Web of Utterance: *Middlemarch.*" *Texas Studies in Literature and Language* 19 (1977): 179–87.

Thayer, Ella Cheever. *Wired Love.* New York: Johnston, 1880.

Thomas, Katie Louise. "Wicked Boys and Unmarried Girls: The Telegraph Worker in Late Nineteenth-Century British Culture." Paper presented at the Northeast Modern Language Association Conference, Pittsburgh, Penn., 17 April, 1999.

"Tricks of the Telegraph." *Punch* 9 (1845): 35.

Twain, Mark. "From the *London Times* of 1904." In *The Science Fiction of Mark Twain,* edited by David Ketterer. Hamden, Conn.: Archon-Shoestring Press, 1984.

———. "Mental Telegraphy." In *The Science Fiction of Mark Twain,* edited by David Ketterer. Hamden, Conn.: Archon-Shoestring Press, 1984.

Tyndall, John. *Fragments of Science: A Series of Detached Essays, Addresses, and Reviews.* 1871. 6th ed. 2 vols. New York: Appleton, 1892.

Veeder, William. "Toxic Mothers, Cultural Criticism: 'In the Cage' and Elsewhere." *Henry James Review* 14 (1993): 264–72.

Volta, Alessandro. *Collezione dell' opere del cavaliere Conte Alessandro Volta.* 3 vols. Firenze: G. Piatti, 1816.

Waldeyer, Wilhelm. "Ueber einige neuere Forschungen im Gebiete der Anatomie des Centralnervensystems." *Deutsche medizinische Wochenschrift* 44 (1891): 1213–18.

Wall, Geoffrey. " 'Different from Writing': Dracula in 1897." *Literature and History* 10.1 (1984): 15–23.

Wellbery, David E. Introduction to *Discourse Networks, 1800/1900,* by Friedrich A. Kittler. Translated by Michael Metteer. Stanford: Stanford University Press, 1990.

"What Are the Nerves?" *The Cornhill Magazine* 5 (1862): 153–66.

Wicke, Jennifer. "Henry James's Second Wave." *Henry James Review* 10.2 (1989): 146–51.

———. "Vampiric Typewriting: *Dracula* and Its Media." *English Literary History* 59 (1992): 467–93.

Wiener, Norbert. *Cybernetics, or Control and Communication in the Animal and the Machine.* 2d ed. Cambridge: MIT Press, 1961.

Williams, Basil. "An Attack on a Telegraph Station in Persia." *Longman's Magazine* 30 (1897): 158–72.

Winter, Alison. "A Calculus of Suffering: Ada Lovelace and the Bodily Constraints on Women's Knowledge in Early Victorian England." In *Science Incarnate: Historical Embodiments of Natural Knowledge,* edited by Christopher Lawrence and Steven Shapin. Chicago: University of Chicago Press, 1998.

———. *Mesmerized: Powers of Mind in Victorian Britain.* Chicago: University of Chicago Press, 1998.

Wright, Walter F. *The Madness of Art: A Study of Henry James.* Lincoln: University of Nebraska Press, 1962.

Yandell, Kay. "The Veritable Maze of Devices: Telegraphic Style and Popular Anxiety in Nineteenth-Century American Imagination." Paper presented at the Northeast Modern Language Association Conference, Pittsburgh, Penn., 17 April, 1999.

Index